蒸烤箱卷 纪念版

中国大锅菜

THE BIG-WOK-MADE
CUISINE OF CHINA

FOOD VOLUME OF STEAM OVEN（COMMEMORATIVE EDITION）

李建国◎主编

首都保健营养美食学会大锅菜烹饪技术专业委员会
北京大地亿仁餐饮管理有限公司
———————————————————— 联合推荐

中国铁道出版社有限公司
CHINA RAILWAY PUBLISHING HOUSE CO., LTD.

图书在版编目（CIP）数据

中国大锅菜：纪念版 . 蒸烤箱卷 / 李建国主编 .—北京：中国铁道
出版社有限公司，2022.5
ISBN 978-7-113-28914-0

Ⅰ . ①中…　Ⅱ . ①李…　Ⅲ . ①中式菜肴－菜谱　Ⅳ . ① TS972.182

中国版本图书馆 CIP 数据核字（2022）第 031969 号

书　　名：**中国大锅菜·蒸烤箱卷（纪念版）**
　　　　　ZHONGGUO DAGUOCAI:ZHENGKAOXIANG JUAN（JINIAN BAN）
作　　者：李建国

责任编辑：王淑艳　　　编辑部电话：(010) 51873022　　　电子邮箱：554890432@qq.com
封面设计：崔丽芳
责任校对：焦桂荣
责任印制：赵星辰

出版发行：中国铁道出版社有限公司（100054，北京市西城区右安门西街 8 号）
网　　址：http://www.tdpress.com
印　　刷：北京盛通印刷股份有限公司
版　　次：2022 年 5 月第 1 版　2022 年 5 月第 1 次印刷
开　　本：889 mm×1 194 mm 1/16　印张：18.75　字数：528 千
书　　号：ISBN 978-7-113-28914-0
定　　价：198.00 元

序言

2021 年底，中国铁道出版社有限公司的王编辑对我说，"中国大锅菜"丛书从 2009 年陆续出版，到现在也有 12 年了，而且一直在热销，她建议我再出一套"中国大锅菜"纪念版。我欣然同意，回顾往昔岁月，虽历经困苦，但终有成就。我就详细讲讲这本书从构思到创作的过程。

2012 年，我退休之后就琢磨继续写书，虽然已经出版一套"中国大锅菜"图书，但我想再增加一些品种，完善此套图书。用万能蒸烤箱做大锅饭是我锁定的写作题材之一，因为这个题材早在 2007 年就有构想。那一年，我去德国 RATIONAL 莱欣诺® 有限公司考察，才真正领悟万能蒸烤箱的魅力。中餐烹饪技法——煎、炒、烹、炸、蒸、煮全能实现！最主要是实现了低油、低糖、低盐，解决人们的"三高"问题，而且把厨师从繁重的劳动中解放出来。我与德国的这家公司谈了写书的想法，当时他们并未在意，而我因工作的原因也没有跟进，就这样一直到了 2015 年。那时，我发现，年轻人不爱干厨房的活，餐饮企业招工也很困难。中餐厨房亟须一场变革，而万能蒸烤箱可以实现这场变革，让年轻人心甘情愿地留在中餐厨房。

这一年，我就着手筹备用万能蒸烤箱做大锅菜，这不是一件容易的事。做事就要做好，否则就不做，这是我一直坚持的原则。怎么做呢，我想了很久，还是得从咱们的"八大菜系"入手。

首先，根据中国的"八大菜系"——川菜、鲁菜、淮扬菜、粤菜、浙菜、闽菜、湘菜、徽菜，制定大锅菜菜谱，提升大锅菜的品质与档次；其次，用万能蒸烤箱做菜，在实践中还未大规模展开。有些企业买了蒸烤箱，只是用来蒸米饭，煮鸡蛋，其他功能并未使用过，实属浪费。我也问过这些厨师，为什么不用它做菜，绝大部分人说没用过，把握不好。因此，我把"八大菜系"的中国烹饪大师们请来，专门用蒸烤箱做菜，其中，也有教学的目的。杜广贝、王志强、侯玉瑞、郑绍武、郑秀生、孙立新、李智东、苏喜斌、王万友、张爱强、王朝晖、赵春源、孙家涛、王素明、徐龙、李加双、林进、林勇、王海东、王连生、王兆志、俞世清、张伟利欣然而来，展现精湛的厨艺，为厨房革命奉献爱心。

其次，用蒸烤箱做菜，温度、湿度、时间都能控制，可批量制作，只要掌握料汁的比例，做团餐是非常合适的，味道与大铁锅做出来的一样，可满足几百人甚至上千人用餐。万能蒸烤箱小的有八层、十层，大的有二十层、四十层（双面），选用哪种类型的万能蒸烤箱根据用餐人数确定。万能蒸烤箱可以实现标准化、批量化制作，味型稳定，且又干净又节省空间。尤其是绿叶类蔬菜，做熟后颜色不变，不同批次的菜肴味道相同。

最后，设想通过"一带一路"，把中国菜传到世界各地，让这些国家和地区的厨师能轻松掌握中国菜的制作方法。

整体构思确定之后，2015 年 11 月底，我在北京朝阳区朝阳路的厨之道美食烹饪教室开工，得到厨之道的鼎立支持，RATIONAL 莱欣诺® 有限公司（北京）的经理赵亮现场操作蒸烤箱，因为有些烹饪大师不太熟悉功能的设置。李君负责摄影，张洋负责文字编写。12 月底就完成前期的制作与拍摄工作，后期图片处理、文字翻译、视频剪辑等工作也在半年内完成。

现在，很多企业都购置了蒸烤箱，说明它已在我国渐渐普及。用现代科学技术呈现中国菜的魅力，对推行中餐标准化，满足高规格的市场需求，打造团膳新天地有着重要意义。

本次出版丛书包括《中国大锅菜·蒸烤箱卷（纪念版）》《中国大锅菜·自助餐副食卷（纪念版）》《中国大锅菜·凉菜卷（纪念版）》《中国大锅菜·热菜卷（纪念版）》《中国大锅菜·主食卷（纪念版）》，在第一版的基础上进行增加或删减，装帧设计为精装，便于读者学习与收藏。

首都保健营养美食学会大锅菜烹饪技术专业委员会

会长　中国烹饪大师　李建国

李智东

高级烹饪技师、国家级考评员。现为北京金手勺餐饮有限公司经理兼行政总厨。

林进

中国烹饪大师、国家级考评员、国家级评审员，现任钓鱼台大酒店厨师长。

林勇

中国烹饪大师、国家级考评员、国家级评审员、人民大会堂副厨师长。

苏喜斌

国家高级烹调技师、餐饮业国家级评委。现任华腾恒逸（北京）国际投资管理有限公司餐战管理分公司运营厨务总顾问。

孙家涛

餐饮业国家一级评委、高级烹调技师、国家级职业技能竞赛裁判员。现任国务院办公厅首长餐厅厨师长。曾参与大型教学系列片中国八大菜系名菜八百例中鲁菜的百菜制作。

孙立新

中国烹饪大师，全国『五一』劳动奖章获得者、中华鲁菜技

王兆志

中国烹饪大师、国家级考评员、国家级评审员。现任钓鱼台十一楼厨师长。

徐龙

中国烹饪大师，团餐与大锅菜专业委员会西餐菜品研发中心主任。现任人民大会堂国宴西餐厨师长、世界御厨协会会员，第六届全国烹饪大赛总决赛第三名获得者，全国技术能手和全国最佳厨师获得者。

张爱强

人民大会堂中华厅经理，国家级高级考评员。

张伟利

中国烹饪大师、国家级考评员、国家级评审员。现任锦江都城酒店行政总厨。

赵春源

中国烹饪大师，团餐与大锅菜专业委员会冷菜及雕刻研发中心主任。高级烹调技师、高级营养配餐师。

郑绍武

中国烹饪大师，团餐与大锅菜专业委员会川菜研发中心主任，

烹饪大师

杜广贝

中国烹饪大师，团餐与大锅菜专业委员会常务副理事长，京菜研发中心主任，中国烹饪协会名厨专业委员会资深委员，国家高级烹调技师，餐饮业国际级评委，国家职业技能鉴定考评员。

侯玉瑞

全国烹饪名师，全国职业技能鉴定教材编委会委员，国家级评委，国家职业技能竞赛裁判员、高级技师、高级考评员，国际美食厨艺联盟专家团队成员。

李加双

中国烹饪大师，国家级考评员、国家级评审员，原总装培训基地副主任。

李建国

中国烹饪大师，团餐与大锅菜专业委员会理事长。1993年全国烹饪大赛金牌获得者，被评为1993年全国百名优秀厨师，国家职业技能鉴定高级考评员，中式烹调高级技师。

艺传承大师。便宜坊集团副总经理兼行政总厨。

王朝辉

便宜坊集团技术研发室、集团产品研发推广事业部副主任，中式烹调高级技师、全国饭店业国家级评委、国家职业技能竞赛裁判员。

王海东

中国烹饪大师，国家高级烹饪技师，北京瑞龙苑宾馆行政总厨。1984年开始从事烹饪工作，多次在全国的烹饪大赛中拿过金牌。先后荣获过「全国最佳厨师」「全国十佳厨师」等称号。

王连生

中国烹饪大师，国家高级烹调技师、国家级考评员。现任北京雪原风情酒楼董事长、总经理。擅长传统鲁菜及东北山珍野味的烹制。

王万友

谭家菜第三代传人。现为国家中餐烹饪高级技师、国家高级营养师、中国烹饪协会裁判委员会委员。

高级烹饪技师、国家级评委，北京市商业服务业中华传统技艺技能大师，川菜制作传承人。

郑秀生

中国烹饪大师，团餐与大锅菜专业委员会淮扬菜研发中心主任，高级烹饪技师，北京饭店行政总厨，获亚洲华人名厨首脑峰会『最高贡献奖』，2010年全国劳动模范等荣誉称号。

俞世清

高级烹调技师、北京烹饪大师、国家级考评员。现任北京聚宝渔港面点技术总监。

王素明

团餐与大锅菜专业委员会面点与烘焙研发中心主任，京西宾馆管理局厨师长、国家中式面点高级技师、餐饮业国家级评委、国家职业技能鉴定高级考评员、北京首届十佳厨师、北京最佳面点师。

王志强

面点烹饪大师、中国烹饪大师、前门饭店面点总厨，国家级考评员。

烹饪大师·厨师漫画

杜广贝　　　　侯玉瑞　　　　李加双　　　　李建国

李智东　　　　林　进　　　　林　勇　　　　苏喜斌

孙家涛　　　　孙立新　　　　王朝辉　　　　王海东

王连生　　　　王万友　　　　王兆志　　　　徐　龙

张爱强　　　　张伟利　　　　赵春源　　　　郑绍武

郑秀生　　　　俞世清　　　　王素明　　　　王志强

目录

菜品名称

北京炒合菜

Name: Stir-fried Different Vegetables by Beijing Style

制作人：杜广贝　　中国烹饪大师

Made by: Guangbei Du　　A Great Master of Chinese Cuisine

主　料　Main Ingredient
猪里脊：1500g　切丝
Pork Tenderloin　1500g　Shred

配　料　Burdening
韭　菜：750g　切段
Chives　750g　Cut
粉　丝：1000g　切段
Vermicelli　1000g　Cut

豆　芽：750g　整根
Bean Sprout　750g　The Original

调　料　Seasoning
清　油　Oil........................180ml
盐　Salt................................20g
酱　油　Soy Sauce..............50ml
胡椒粉　Ground Pepper.........10g
料　酒　Cooking Wine.........30ml

姜　末　Minced Ginger..........30g
葱　末　Minced Scallion.........30g

备　注　Tips
使用炒锅烧制此菜，最讲究锅气，而使用万能蒸烤箱则将香味储存在料汁里。
The Universal Steam Oven can keep the fragrance into the vegetables.

中国大锅菜

蒸烤箱卷（纪念版）

The Big-Wok-Made Cuisine of China, Food Volume of Steam Oven（Commemorative Edition）

制作方法

❶ 首先炒制料汁，锅热下油180ml，将里脊丝放入，将水分煸出，待肉丝炒熟，加葱、姜末各30g，料酒30ml，酱油50ml，胡椒粉10g，盐20g，倒入鸡汤750ml，料汁即成。在料汁中加入粉丝，让粉丝上色，继续烧至粉丝熟透。

❷ 韭菜用少许油搅拌，将韭菜和豆芽分别放入万能蒸烤箱飞水，选择『蒸制蔬菜』模式，韭菜时长30秒，豆芽时长2分钟。

❸ 将飞水后的韭菜和豆芽盛入布菲盒中，倒入烧制好的粉丝料汁，搅拌均匀即可盛盘。

WORKING PROCESS

1. Make the sauce first, pour the oil 180ml into a heated wok, deep-fry the pork tenderloin till the meat becomes dry, add minced scallion 30g, minced ginger 30g, cooking wine 30ml, soy sauce 50ml, ground pepper 10g, salt 20g, pour the chicken soup 750ml, the sauce is ready. Add vermicelli into the soup, boil till the soup boiled.

2. Stir chives with a little oil, steam the chives and bean sprout via the Universal Steam Oven, select "Steam Vegetable" mode, 30 seconds for chives and 2 minutes for bean sprout.

3. Put the steam chives and bean sprout into a buffet pot, pour the sauce and stir evenly.

中国大锅菜

菜品名称 · 北京炒合菜

Name: Stir-fried Different Vegetables by Beijing Style

菜品特点

特色 炒合菜是一道北京的传统菜，顾名思义，就是将几种食材放一起炒制，其制作方法简单，味道却不简单。它有两个特点：首先是时令性强，俗语云：『正月葱，二月韭』，以春天的韭菜制作此菜最好不过；其次搭配春饼卷食，更有一番滋味。

品味 这是一道十分鲜美的菜肴，咸鲜适口，韭菜在其中发挥了至关重要的作用。尤其是春天的韭菜，那种香气是每一名食客都无法拒绝的。每一种食材都发挥了自己的味道，组合在一起合而不散，香气诱人。整道菜看上去颜色略深，酱香之色浓郁。

品相 用万能蒸烤箱制作此菜，可以很好地避免锅内汤汁过多的情况，易出汤的韭菜和豆芽在飞水时都可以将流出的汤汁滤掉，搅拌在一起能充分吸收料汁的香味。

营养价值 这道菜食材种类很多，在营养价值方面也是各显神通。整道菜清淡可口，营养丰富。猪肉含有丰富的优质蛋白质和人体必需的脂肪酸，韭菜为辛温补阳之品，含有一定量的锌元素，能温补肝肾，豆芽富含大量的维生素C，可以有效预防坏血病，具有清除血管壁中的胆固醇和脂肪的堆积，防止心血管病变的作用；粉条里富含碳水化合物、膳食纤维、矿物质等。

菜品名称

葱油木耳白菜

Name: Stir-fried Chinese Cabbage and Black Fungus with Scallion Oil

制作人：杜广贝　　中国烹饪大师

Made by: Guangbei Du　　A Great Master of Chinese Cuisine

主 料 Main Ingredient	配 料 Burdening	调 料 Seasoning
白 菜：1500g 切片	木 耳：500g 改刀	葱 油 Scallion oil...............30ml
Chinese cabbage 1500g Sliced	Black fungus 500g Sliced	盐 Salt 20g
	胡萝卜：500g 切片	
	Carrot 500g Sliced	

中国大锅菜

菜品名称 · 葱油木耳白菜
Name: Stir-fried Chinese Cabbage and Black Fungus with Scallion Oil

制 作 方 法

将白菜、木耳、胡萝卜倒入盆中，加盐20g，葱油30ml，搅拌均匀。然后放入万能蒸烤箱炒制，选择『单点分层煎烤』模式。时长3分30秒，即可出锅。

WORKING PROCESS

Put the Chinese cabbage, black fungus, carrot into a basin, add salt 20g, scallion oil 30ml, stir evenly, stir-fry at the Universal Steam Oven, select "Single Point Stratified Bake" mode for 3.5 minutes, then the dish's done.

中国大锅菜

蒸烤箱卷（纪念版）

The Big-Wok-Made Cuisine of China, Food Volume of Steam Oven (Commemorative Edition)

菜品特点

特色 这是一道经典家常菜，将白菜和木耳炒在一起，以葱花炝锅，香气四溢，广受喜爱。

品味 葱是中餐烹饪的主要调味品之一，这道菜调味料十分简单，突出的是食材本身的味道，葱香扑鼻，而白菜清甜可口，有种淡淡的香气，两者结合，辅以木耳，虽然菜色清淡，但口味却并不寡淡。

品相 烹饪中无须添加酱油等调味品，以展现食材本色为佳，黑白之间，略有一点胡萝卜的橙红。

营养价值 白菜营养价值丰富，是我国居民冬季餐饮中最主要的蔬菜之一。白菜含有多种营养物质，是人体生理活动所必需的维生素、无机盐及食用纤维素的重要来源，并含有丰富的钙，是预防癌症、糖尿病和肥胖症的健康食品。木耳中铁的含量极为丰富，故常吃木耳能养血驻颜，令人肌肤红润，容光焕发，并可防治缺铁性贫血。木耳中的胶质可把残留在人体消化系统内的灰尘、杂质吸附集中起来排出体外，从而起到清胃涤肠的作用。

菜品名称

宫保羊肉

Name: Kung Pao Mutton

制作人：杜广贝　　中国烹饪大师

Made by: Guangbei Du　　A Great Master of Chinese Cuisine

主　料 Main Ingredient
羊　肉：2000g　切丁
Mutton　2000g　Pieced

配　料 Burdening
青、红椒：1000g　切片
Green/Red Bell
Pepper　1000g　Sliced

花生豆：500g　去皮
Peanut　500g　Peeled

调　料 Seasoning
清　油 Oil......................150ml
盐 Salt 40g
酱　油 Soy Sauce 140ml
白 糖 Sugar......................60g

料　酒 Cooking Wine 220ml
胡椒粉 Ground Pepper 10g
淀　粉 Starch.......................80g
鸡　蛋 Egg.................2 个 (pcs)
花　椒 Chinese Prickly Ash ..10g
干辣椒 Dried Chili 30g
醋 Vinegar........................150ml
料　油 Spicing Oil 80ml

中国大锅菜

The Big-Wok-Made Cuisine of China, Food Volume of Steam Oven（Commemorative Edition）

蒸烤箱卷（纪念版）

制作方法

❶ 将 2 个鸡蛋清加 80g 淀粉制成鸡蛋糊备用。

❷ 首先将羊肉上浆腌制，羊肉入盆，加料酒 40ml，胡椒粉 10g，酱油 40ml，盐 20g，倒入水 30ml，抓匀，增加羊肉的底味，再倒入鸡蛋糊，加料油 10ml，搅拌均匀。挂糊上浆，腌制 15 分钟。

❸ 烧制宫保汁，锅热下油 150ml，加入花椒 10g，煸香后捞出，倒入干辣椒段 30g，煸香后捞出备用，倒入水 1.2L，锅开后加糖 60g，料酒 180ml，酱油 100ml，醋 150ml，盐 20g，烧开后加料油 70ml，料汁即成。

❹ 烤盘刷底油，将腌制好的羊肉码入盘中，放入万能蒸烤箱滑油，选择『单点分层炙烤』模式，时长 3 分 30 秒。

❺ 将青、红椒片倒入滑油后的羊肉中，均匀浇上宫保汁，放入万能蒸烤箱烹制入味，选择『单点分层煎烤』模式，时长 3 分钟。

❻ 出锅后，撒上备用的辣椒段和花生豆，即可盛盘。

WORKING PROCESS

1. Mix the egg-white with starch 80g to make egg mash for later use.

2. Coating and marinating the mutton. Put the mutton into basin, pour cooking wine 40ml, ground pepper 10g, soy sauce 40ml, salt 20g, pour water 30ml, well stir and knead, to flavor the mutton, add spicing oil 10ml, well stir and coat it. Marinate for 15 minutes.

3. Make Kung Pao sauce, pour oil 150ml into a heated wok, add Chinese prickly ash 10g, stir-fry and take out, pour water 1.2L, add sugar 60g while the water boiled, cooking wine 180ml, soy sauce 100ml, vinegar 150ml, salt 20g, boil the sauce and add spicing oil 70ml, the sauce is ready.

4. Brush the oil at the bottom of the baking pan, put the marinated mutton to the baking pan, stir-fry at the Universal Steam Oven, select "Single Point Stratified Grill" mode for 3.5 minutes.

5. Sprinkle the green/red bell pepper into the mutton, pour the sauce evenly, bake in the Universal Steam Oven, select "Single Point Stratified Bake" mode for 3 minutes.

6. The last step, sprinkle the peanuts and chili then the dish is done.

中国大锅菜

菜品名称·宫保羊肉
Name: Kung Pao Mutton

菜品特点

特色 此菜由传统名菜「宫保鸡丁」演变而来，羊肉经过腌制后十分软嫩，并且其本身便带有一股诱人的香味，相比宫保鸡丁，有「青出于蓝，而胜于蓝」的感觉。

品味 宫保汁酸甜适口，甜中有辣，属于复合味型；羊肉香气浓郁，口感软嫩，就花生而食，浓郁的香味为这道菜更添光彩。由于羊肉本身有膻味，需要更多的调味料，所以相比宫保鸡丁口味更重些。

品相 此菜料汁颜色红亮而酱色略深，羊肉上裹满料汁，配以青红椒，色泽鲜艳。

营养价值 羊肉含有丰富的蛋白质，较猪肉、牛肉高，而且比猪肉和牛肉的脂肪、胆固醇含量都要少。羊肉肉质细嫩，容易消化吸收，多吃羊肉有助于提高身体免疫力。中医上讲羊肉性热，味甘，能助元阳，补精血，疗肺虚，益劳损，适宜于冬季进补，是补阳的佳品。

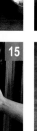

菜品名称

萝卜汆鱼片

Name: Qucik-Boiled Long Li Fish with Radish

制作人：杜广贝　　中国烹饪大师

Made by: Guangbei Du　　A Great Master of Chinese Cuisine

主　料　Main Ingredient

龙俐鱼：2000g　切片

Long Li Fish　2000g　Sliced

配　料　Burdening

白萝卜：1500g　切片

White radish　1500g　Sliced

胡萝卜：500g　切片

Carrot　500g　Sliced

调　料　Seasoning

鸡　蛋　Egg ....3 个 (pcs)

盐　Salt 30g

淀　粉　Starch 50g

胡椒粉　Ground Pepper 5g

料　酒　Cooking Wine 30ml

葱　片　Sliced Scallion...........20g

姜　片　Sliced Ginger 20g

备　注　Tips

加鸡蛋糊上浆时，糊不要太稠。

Don't make the egg-mash too thick.

制作方法

❶ 将3个鸡蛋清打入小盆，加淀粉50g拌匀，制成鸡蛋糊备用。

❷ 首先将鱼片腌制上浆，鱼片入盆，加葱片20g，姜片20g，料酒30ml，盐15g，胡椒粉5g，拌匀后加上鸡蛋糊，腌制15分钟。

❸ 将白萝卜和胡萝卜放入深烤盘，倒入水，没过菜品，放入万能蒸烤箱加热，选择『单点分层蒸煮』模式，时长20分钟。

❹ 白萝卜和胡萝卜出锅后，均匀撒上盐15g，摆上腌制好的鱼片，放入万能蒸烤箱烹制入味，选择『单点分层蒸煮』模式，时长4分钟，即可出锅。

WORKING PROCESS

1. Break 3 eggs in a basin, mix the starch 50g, make the egg-mash for later use.

2. Coating the fish. Put the fish slices in the basin, add sliced scallion 20g, sliced ginger 20g, cooking wine 30ml, salt 15g, ground pepper 5g, well stir then pour the egg-mash, marinate for 15 minutes.

3. Get together the white radish and carrot into baking pan, pour water till cover them, bake them by using the Universal Steam Oven, select "Single Point Stratified Steam" mode for 20 minutes.

4. Sprinkle salt 15g after baking, put the marinated fish slices, steam it via the Universal Steam Oven, select "Single Point Stratified Steam" mode for 4 minutes. The dish is done.

中国大锅菜

蒸烤箱卷（纪念版）

The Big-Wok-Made Cuisine of China, Food Volume of Steam Oven（Commemorative Edition）

菜品特点

特色 『汆』是中国传统烹饪方法之一，将食材放入汤中煮一下，随即取出。这种烹调方法可以防止食物本身变柴、变硬，亦可减少食物高温烹调而流失的养分。这道菜就是将腌制好的鱼片在萝卜丝汤中汆制而成。

品味 鱼片经过腌制后去腥取鲜，十分入味，鲜美无比，用汆制的烹调方法，进一步保留了食材的鲜香，口感也十分软嫩。白萝卜经过长时间的炖煮后辛辣味尽除，香味蹿满整个汤底。整道菜口味十分清淡，低油、低盐。萝卜也是冬季滋补佳品，吃完鱼肉后，盛一碗汤喝，温暖可口。

品相 这道菜十分清淡，品相也呈现出素雅之色，鱼片和白萝卜皆为白色，点缀以胡萝卜，白里带红，让菜品富有生机。

营养价值 这道菜清淡可口，是一道养生之品。制作此菜对鱼的种类选择不甚挑剔，以味道更为鲜美的海鱼为佳，本菜选用的是龙俐鱼。龙俐鱼含有较高的不饱和脂肪酸，蛋白质容易消化吸收，其肌肉细嫩，口感爽滑，鱼肉久煮不老，无腥味和异味。白萝卜含丰富的维生素C和微量元素锌，有助于增强机体的免疫功能，提高抗病能力。

东安仔鸡

Name: Dong An Young Chicken

制作人：侯玉瑞　　中国烹饪大师

Made by: Yurui Hou　　A Great Master of Chinese Cuisine

主　料　Main Ingredient
鸡腿肉：1500g　切条
Drumstick　1500g　Stripped

配　料　Burdening
青、红椒：250g　切片
Green/red Bell Pepper 250g Sliced
洋　葱：250g　切丝
Onion　250g　Shredded

调　料　Seasoning
干辣椒 Dried Chilli................15g

盐 Salt 15g
醋 Vinegar........................320ml
料　酒 Cooking Wine 20ml
香　油 Sesame Oil 30ml
淀　粉 Starch......................40g
鸡　蛋 Egg.................1 个（pcs)
胡椒粉 Ground Pepper...........6g
姜　丝 Shredded Ginger 150g

备　注　Tips
1. 腌制时，盐的量是肉的 1% 左右；
放辣椒碎和醋是为了事先将底味上

足，更加入味。
2. 炒汁时要二次投姜，汁中有生姜
味、有熟姜味。

1. The proportion of salt should be
1% of the meat. The chopped chili
and vinegar help to marinate the
chicken as well.
2. It needs adding girger twice during
cooking to enhance the taste.

中国大锅菜

蒸烤箱卷（纪念版）

The Big-Wok-Made Cuisine of China, Food Volume of Steam Oven（Commemorative Edition）

制作方法

❶ 将10g干辣椒放入干锅中煸炒，将辣椒煸脆，潮气煸干，颜色变深，并压碎制成辣椒碎备用。将20g淀粉制成水淀粉备用。

❷ 鸡条入盆腌制，加料酒20ml，胡椒粉6g，盐15g，1个鸡蛋，辣椒碎10g，淀粉20g，醋20ml，姜丝50g，揉拌均匀，腌制15分钟。

❸ 将腌制好的鸡条滑油，烤盘刷香油，码入鸡条，选择『单点分层炙烤』模式，时长1分钟。

❹ 炒制料汁，锅热下香油20ml，加干辣椒5g，煸香，加姜丝60g，醋300ml，烹制出香气，二次投姜40g，加水淀粉勾芡，倒入青红椒段和洋葱丝，继续煸炒少许，料汁即成。

❺ 将料汁倒入滑油后的鸡条，放入万能蒸烤箱烹制入味，选择『单点分层炙烤－薄－1号色』，时长5分钟，即可出锅。

WORKING PROCESS

1. Stir-fry the dried chili 10g in a wok till it turns crisp, gets rid of the moist, turns its color to be heavy, pulverize it for later use. Turn starch 20g into water-starch mash for later use.

2. Marinate the drumstick in a basin, add cooking wine 20ml, ground pepper 6g, salt 15g, 1 egg, pulverized dried chili 10g, starch 20g, vinegar 20ml, shredded ginger 50g, well stir it and marinate for 15 minutes.

3. Quick stir-fry the drumstick, brush sesame oil on the baking pan, place the stripped drumstick, select "Single Point Stratified Grill" mode for 1 minute.

4. Make the sauce, pour sesame oil 20ml into a heated wok, add dried chili 5g, stir-fry it till fragrance and add stripped ginger 60g, vinegar 300ml, cook till fragrance, then add ginger 40g, thicken the soup with water-starch mash, add stripped onion and sliced green/red bell pepper, continue to stir-fry and the sauce is done.

5. Dress the sauce into the quick stir-fried stripped drumsticks, cook by using the Universal Steam Oven, select "Single Point Stratified Grill" for 5 minutes. Then the dish is done.

中国大锅菜

菜品名称·东安仔鸡

Name: Dong An Young Chicken

菜品特点

特色　这是一道湘菜名菜，起源于湖南永州东安县，原名『醋鸡』。据传，北伐胜利后，国民革命军第八军军长唐生智在南京曲园酒家设宴，席间特别吩咐上一道醋鸡，众人品尝后赞不绝口，问及来历，唐军长觉醋鸡一名不雅，便以家乡之名为此菜定名为『东安仔鸡』。1972年，美国总统尼克松访华期间，毛泽东设宴款待，席间就有此菜。尼克松十分称赞，回国后亦多次与人谈及。用万能蒸烤箱制作此菜是一大探索。

品味　这道菜香气四溢，整体给人以满口鲜香的味觉享受，仔细品味，又能感觉到每一种食材的味道，搭配在一起，真是相得益彰。

鸡肉软嫩，醋经过烹调后酸味淡去，留下醇香之味，制作此菜以米醋为佳，为菜品奠定了浓香的基调，由于烹调中二次投姜，菜品既有生姜味，又有熟姜味，香中有鲜。

品相　鸡肉黄亮，配以青、红椒段，色彩鲜艳夺目，从透亮的芡汁就可以感受到鸡肉的软嫩和菜肴的鲜香。

营养价值　鸡肉性平、温、味甘，入脾、胃经，可益气、补精、添髓。而鸡翅中的胶原蛋白含量更加丰富，对于保持皮肤光泽、增强皮肤弹性均有好处。

冬瓜汆丸子

Name: Quick-Boiled Pork Balls with Wax Gourd

制作人：侯玉瑞　　中国烹饪大师

Made by: Yurui Hou　　A Great Master of Chinese Cuisine

主　料　Main Ingredient

冬　瓜：2500g　切片
Wax Gourd　2500g　Sliced
五花肉：1000g　切末
Pork Belly　1000g　Minced

配　料　Burdening

鸡　蛋：1个
Egg　1个 (pcs)
香菜叶：20g
Caraway Leaves　20g

调　料　Seasoning

香　油　Sesame Oil 20ml
胡椒粉　Ground Pepper 6g
淀　粉　Starch 20g
盐　Salt 12g
姜　末　Minced Ginger 30g
葱　末　Minced Scallion 20g

备　注　Tips

1. 拌肉馅时，要根据实际情况往肉馅中打入尽可能多的水。
2. 一定要将丸子放入凉水中，一起倒入烤盘。
3. 丸子变色即熟，不要蒸制过长时间，否则不够软嫩。

1. More water is need to make the meat stuffing.
2. Put the meatballs into cool water and then pour them into baking pan.
3. Don't steam the meatballs too long time.

17

中国大锅菜

菜品名称 · 冬瓜氽丸子

Name: Quick-Boiled Pork Balls with Wax Gourd

制 作 方 法

❶ 首先将冬瓜码入烤盘，倒入开水 1.5L，加胡椒粉 3g，盐 4g，姜末 10g，拌匀后倒入烤盘，蒸制冬瓜，选择『单点分层蒸煮』模式，时长 3 分钟。

❷ 将五花肉馅倒入盆中，准备 500ml 水揉制。将水打入肉馅，加胡椒粉 3g，盐 8g，打入 1 个鸡蛋，搅拌均匀，加葱末、姜末各 20g，干淀粉 20g，拌匀，倒入香油 20ml，拌匀后馅料即成。

❸ 取净盆，倒入凉水 1L，将馅料制成丸子，放入水中。冬瓜蒸好后，将盆内的丸子和水倒入冬瓜中，放入万能蒸烤箱，烹制成熟，选择『单点分层蒸煮』模式，时长 6 分钟，水开后肉丸变色即熟。出锅后撒上香菜叶 20g，即可盛盘。

WORKING PROCESS

1. Place the wax gourd into the baking pan, get a clean basin, pour boiled water 1.5L, add ground pepper 3g, salt 4g, minced ginger 10g, well stir and mix in baking pan. Steam the wax gourd, select "Single Point Stratified Steam" mode for 3 minutes.

2. Place the pork belly meat-stuffing into a basin, add water 500ml, knead it and well stir. Add ground pepper 3g, salt 8g, 1 egg, minced scallion 20g, minced ginger 20g, starch 20g, well stir and pour sesame oil 20ml, well stir and the meat-stuffing is done.

3. Get a new basin, pour cool water 1L into it, make the meatballs, place into water. Steam wax gourd, mix the meatballs, water with the wax gourd, cook by using the Universal Steam Oven, select "Single Point Stratified Steam" mode for 6 minutes, cook till the water boiled. Then sprinkle caraway leaves 20g, the dish is done.

中国大锅菜

The Big-Wok-Made Cuisine of China, Food Volume of Steam Oven（Commemorative Edition）

蒸烤箱卷（纪念版）

菜品特点

特色 这是一道家常菜。冬天时，与母亲相对而坐，品尝砂锅冬瓜余丸子，最后把汤都喝干净，满满都是幸福的回忆。这道菜食材普通，制作方法简单，营养十分丰富，口味清淡，但味道却丝毫不逊色。

品味 五花肉香味浓厚，由于打入了很多水，口感软嫩。葱、姜提升了整个菜品的鲜味，入口鲜香，冬瓜有着独特的清香，对于肉味是很好的调剂。

品相 肉丸发白，略透出一点红色，冬瓜翠白，汤汁清淡，冒着腾腾热气，十分暖人。

营养价值 五花肉含有丰富的优质蛋白质和必需的脂肪酸，并提供血红素（有机铁）和促进铁吸收的半胱氨酸，能改善缺铁性贫血；具有补肾养血，滋阴润燥的功效。冬瓜含维生素C较多，且钾盐含量高，钠盐含量较低，高血压、肾脏病、浮肿病等患者食之，可达到消肿而不伤正气的作用。

荷叶粉蒸肉

Name: Steamed Pork with Rice Flour in Lotus Leaves

制作人：侯玉瑞　　中国烹饪大师

Made by: Yurui Hou   A Great Master of Chinese Cuisine

主　料　Main Ingredient
五花肉：1000g　切大片
Pork Belly　1000g　Sliced

配　料　Burdening
豆腐丝：1000g　切丝
Dried Bean Curd　1000g　Shredded

调　料　Seasoning
米　粉　Rice Flour 450g
盐　Salt 5g
酱　油　Soy Sauce 10ml
香　油　Sesame Oil 10ml
蚝　油　Oyster Sauce 20ml
料　酒　Cooking Wine 15ml
姜　末　Minced Ginger 10g
甜面酱　Sweet Bean Paste 10g

备　注　Tips
1. 拌米粉时，一定要将肉片拌散，不要两片肉粘连在一起。

2. 荷叶要盖严，不要让味道跑出过多。

1. Make sure to stir the rice flour and the meat evenly.

2. To cover the dish strictly by lotus leaves.

中国大锅菜

制作方法

① 首先将荷叶铺在烤盘上，将豆腐丝均匀码在荷叶上，然后给豆腐丝上底味，加盐5g、香油10ml、姜末10g，拌匀即可。

② 将五花肉片入盆腌制，加料酒15ml、水10ml、蚝油20ml、酱油10ml、甜面酱10g，加调料过程中要不停搅拌，让调料与食材充分接触。再加米粉450g，将肉片裹上米粉。

③ 将裹上米粉的肉片码在豆腐丝上，盖上荷叶。如果荷叶不够大，就另取一张荷叶盖上。放入万能蒸烤箱蒸制，选择『单点分层蒸煮』模式，时长40分钟，即可出锅。

WORKING PROCESS

1. Pave the lotus leaves on the baking pan, then place the shredded dried bean curd evenly on it, then add salt 5g, sesame oil 10ml, minced ginger 10g, well stir.

2. Marinated the pork belly into a basin, add cooking wine 15ml, water 10ml, oyster sauce 20ml, soy sauce 10ml, sweet bean paste 10g, continue to stir, then add rice flour 450g, coat the meat with the rice flour.

3. Put the coated sliced pork on the shredded dried bean curd, cover by lotus leaves. Steam it by using the Universal Steam Oven, select "Single Point Stratified Steam" mode for 40 minutes. Then the dish is done.

中国大锅菜

菜品名称·荷叶粉蒸肉

Name: Steamed Pork with Rice Flour in Lotus Leaves

菜品特点

特色 这是一道湘菜，但却起源于江西。袁枚随园食单录有该菜记载：『用精肥参半之肉，炒米粉黄色，拌面酱蒸之，下用白菜作垫，熟时不但肉美，菜亦美。以不见水，故味独全。江西人菜也。』明初『江西填湖广』，是我国历史上规模较大移民潮之一，今天许多湘菜都可以在江西找到根源。传到湖南后逐渐用荷叶包裹，因为这样便于携带和取出食用，无须碗筷，在劳作歇息时充饥。

品味 采用荷叶也许是无心插柳之举，却诞生了一道经典菜肴，肉软嫩鲜美，裹上米粉又使菜品增加了谷物的香甜，荷叶更是点睛之笔，它沁人心脾的清香扑鼻而来，又缓解了五花肉的肥腻，一举两得。垫菜的选择多种多样，江西用白菜，本菜谱用豆腐丝，二者均能在吸收油脂的同时使本身的口味增色不少。

品相 打开荷叶，香气腾腾的，绝对是一大诱惑，米粉肉呈现出较浅的酱色，盖在嫩白的豆腐丝上，软糯、醇香。

营养价值 据本草纲目记载：『荷叶性味苦、平，能清热、解毒，散瘀，治暑热等。』所以荷叶粉蒸肉除了营养丰富，味道鲜美外，还有一定的药用效果。五花肉含有丰富的优质蛋白质和必需的脂肪酸，并提供血红素（有机铁）和促进铁吸收的半胱氨酸，能改善缺铁性贫血；具有补肾养血，滋阴润燥的功效。

酿 苦 瓜

Name: Stuffed Balsam Pear

制作人：侯玉瑞　　中国烹饪大师

Made by: Yurui Hou　　A Great Master of Chinese Cuisine

主 料　Main Ingredient
苦 瓜：750g　切段
Balsam pear　750g　Cut

配 料　Burdening
猪五花肉：500g　切末
Pork Belly　500g　Minced
韭 菜：250g　切末
Chinese Chives　250g　Chopped

调 料　Seasoning
清 油　Oil.............................20ml
盐 Salt....................................10g
酱 油　Soy Sauce 5ml
胡椒粉　Ground Pepper...........3g
白 糖　Sugar........................10g
淀 粉　Starch.......................10g
美极鸡汁　Maggi Chicken
　　　　　Sauce..................15ml
姜 末　Minced Ginger 15g

葱 末　Minced Scallion.........15g

备 注　Tips
往苦瓜内放馅时，馅料要冒出来一
些，这样不会因为受热收缩而无法挂
在苦瓜上。
More stuffing should be stuffed
into the balsam pear to avoid the
shrinking of the meat.

中国大锅菜

菜品名称·酿 苦 瓜
Name: Stuffed Balsam Pear

制作方法

❶ 将五花肉末和韭菜倒入盆中，加葱末 5g，姜末 5g，酱油 5ml，盐 10g，胡椒粉 3g，拌成肉馅。

❷ 将肉馅塞入苦瓜段中，均匀码入烤盘，选择『单点分层蒸煮』模式，时长 7 分钟。

❸ 蒸苦瓜时炒制料汁，锅热下油 20ml，加葱、姜末各 10g 爆香，加糖 10g，倒入 100ml 水，美极鸡汁 15ml，锅开后加 10g 淀粉制成的水淀粉勾芡，料汁即成。

❹ 苦瓜馅蒸好后浇上料汁，菜品即成。

WORKING PROCESS

1. Mince the pork belly and mix it with Chinese chives, add minced scallion 5g, minced ginger 5g, soy sauce 5ml, salt 10g, ground pepper 3g, make the stuffing.

2. Stuff the balsam pear, then place them into the baking pan, select "Single Point Stratified Steam" mode for 7 minutes.

3. Make the sauce at the same time. Pour oil 20ml into a heated wok, then add minced scallion 10g, minced ginger 10g, quick stir-fry till fragrance, then add sugar 10g, water 100ml, maggi chicken sauce 15ml, pour starchy mash to thicken the soup after the soup boiled, the sauce is done.

4. Pour the sauce on the balsam pear, the dish is done.

中国大锅菜

The Big-Wok-Made Cuisine of China, Food Volume of Steam Oven（Commemorative Edition）

蒸烤箱卷（纪念版）

东坡肉　　　　酸辣藕丁

菜品特点

特色　酿的繁体字写作『釀』，是形声字，从酉从襄，襄亦平声，『襄』意为『包裹』『包容』。酿苦瓜就是将苦瓜掏空，包裹肉馅。这道菜是客家菜的经典菜，类似菜品还有酿豆腐、酿茄子，因为客家人迁到南方地区后无面粉包饺子，便制作酿菜来代替。客家菜中有许多烹饪方法依然保留着古代中原地区的传统，比如食物多蒸煮而少炸烤，口味上注重鲜香，讲究原味。

品味　此菜采用蒸制的方法烹制，最大程度保留了食材本身的味道，肉馅香味浓厚，苦瓜淡苦清香，二者口味在冲突中得以融合，一起将鲜味体现得淋漓尽致。

品相　这是一道手工菜，造型精致，摆盘美观，芡汁亮透，用万能蒸烤箱蒸制的苦瓜色泽尤为鲜绿，令人赏心悦目。

营养价值　苦瓜中含有丰富的维生素C和苦味甙，苦味素，苦瓜素被誉为『脂肪杀手』，能减少脂肪和多糖的摄取。中医上讲苦瓜具有清热消暑、养血益气、补肾健脾、滋肝明目的功效。猪肉含有丰富的优质蛋白质和必需的脂肪酸，并提供血红素（有机铁）和促进铁吸收的半胱氨酸，能改善缺铁性贫血。

豆瓣鱼块

Name: Deep-fried Fish with Pixian Chili Bean Paste

制作人：李加双　　中国烹饪大师

Made by: Jiashuang Li　　A Great Master of Chinese Cuisine

主　料　Main Ingredient
草　鱼：2250g　切块
Grass Carp　2250g　Cut

调　料　Seasoning
清　油 Oil.........................180ml
盐　Salt 15g
酱　油 Soy Sauce 20ml
郫县豆瓣酱 Pixian Chili Bean

Paste 150g
料　酒 Cooking Wine 80ml
醋 Vinegar.......................180ml
白　糖 Sugar........................80g
香葱碎 Chopped Chive.........30g
淀　粉 Starch.......................70g
姜 Ginger 130g
葱 Scallion 130g
蒜　末 Minced Garlic..........300g

备　注　Tips
1. 鱼腌制的时间不宜过长，但太短
又不能入味，以 15 分钟为宜。
2. 料汁一定要多，否则无法让鱼块
充分入味。
1. The marinating of the fish should
be controlled at about 15 minutes.
2. The fish can be marinated fast
with sufficient sauce.

中国大锅菜

蒸烤箱卷（纪念版）

制作方法

❶ 将草鱼入盆腌制，加料酒 30ml、葱段 30g、姜片 30g、盐 5g，拌匀后腌制 10 到 15 分钟。将 70g 淀粉制成水淀粉备用。

❷ 烤盘刷油，鱼块上也刷少许明油，将腌制好的鱼块放入万能蒸烤箱炸制，选择『单点分层炙烤』模式，时长 10 分钟。

❸ 炸鱼时炒制料汁，锅热下油 180ml，油热加郫县豆瓣酱 150g 炒出香气。加蒜末 300g、姜末 100g、葱花 100g，爆香后倒入水 1.7L，锅开倒入料酒 50ml，酱油 20ml、糖 80g、盐 10g、醋 180ml，一边搅拌一边倒入水淀粉，芡汁即成。

❹ 将芡汁浇在炸制好的鱼上，放入万能蒸烤箱烹制入味，选择『单点分层煎烤』模式，时长 5 分钟。出锅后撒上香葱碎，菜品即成。

WORKING PROCESS

1. Marinate the grass carp in a basin. Add cooking wine 30ml, scallion 30g, sliced ginger 30g, salt 5g, well stir and marinate for 10 to 15 minutes. Turn the starch 70g into water-starch mash for later use.

2. Brush oil on the baking pan and the fish. Bake the marinated fish by using the Universal Steam Oven, select "Single Point Stratified Grill" mode for 10 minutes.

3. Make the sauce at the same time. Pour oil 180ml into a heated wok, add Pixian chili bean paste 150g, stir-fry till fragrance. Add minced garlic 300g, minced ginger 100g, chopped scallion 100g, deep-fry and pour water 1.7L, pour cooking wine 50ml after the water boiled, soy sauce 20ml, sugar 80g, salt 10g, vinegar 180ml, stir it and add water-starch mash to thicken the soup. Then the sauce is done.

4. Coat the sauce on the deep-fried the fish, then bake it via using the Universal Steam Oven, select "Single Point Stratified Grill" mode for 5 minutes. Sprinkle chopped chive afterwards, the dish is done.

中国大锅菜

菜品名称·豆瓣鱼块
Name: Deep-fried Fish with Pixian Chili Bean Paste

菜品特点

特色 豆瓣鱼块是四川地区一道家常菜，鱼选用最常见的草鱼即可。豆瓣指的就是豆瓣酱，以郫县所产最为著名。豆瓣酱被誉为「川菜之魂」，是四川老百姓餐桌上的常客，很多家庭都会根据口味自己腌制，味道不同，甚至会出现「一菜百味」的现象。

品味 鱼为至鲜之物，经过腌制过后加油烹炸，既去除了腥味，又保留了鱼肉的鲜嫩。用豆瓣酱炒制的料汁，鲜辣之中又有回味的香甜，辣而不燥，红油润泽，豆瓣香辣，料汁的香辣与鱼肉的鲜嫩充分结合，形成了令人久久难以忘怀的鲜香。

品相 这道菜料汁浓稠，色泽红亮，鱼肉的嫩白被隐藏在红油之中，菜色充满了香辣的诱惑。

营养价值 草鱼含有丰富的不饱和脂肪酸，对血液循环有利，是心血管病人的良好食物；它还含有丰富的硒元素，经常食用有抗衰老、养颜的功效，而且对肿瘤也有一定的防治作用；此外，对于身体瘦弱、食欲不振的人来说，草鱼肉嫩而不腻，可以开胃、滋补。

菜品名称

黑白豆腐

Name: Baked Black and White Tofu

制作人：李加双　　中国烹饪大师

Made by: Jiashuang Li　　A Great Master of Chinese Cuisine

主　料 **Main Ingredient**
豆　腐：750g　切块
Tofu　750g　Cut
猪血块：750g　切块
Pig Blood Block　750g　Cut

配　料 **Burdening**
牛　肉：250g　切末
Beef　250g　Minced

调　料 **Seasoning**
清　油　Oil 100ml
盐　Salt 5g
酱　油　Soy Sauce 15ml
白　糖　Sugar........................10g
豆　豉　Fermented Soya
　　　　Beans..........................10g
料　酒　Cooking Wine 30ml
淀　粉　Starch.......................30g
郫县豆瓣酱 Pixian Chili Bean

Paste 100g
葱　末　Minced Scallion.........20g
姜　末　Minced Ginger 20g

备　注 **Tips**
由于豆腐出水较多，所以茨汁要相对
浓稠些。
The starchy sauce should be a little
bit thicker because tofu has water.

中国大锅菜

菜品名称 · 黑白豆腐
Name: Baked Black and White Tofu

制作方法

❶ 将豆腐加盐5g拌匀，倒入深烤盘，再倒入没过豆腐的水，放入万能蒸烤箱焯水，选择『单点分层蒸煮』模式，时长5分钟。将30g淀粉制成水淀粉备用。

❷ 飞水时炒制料汁，锅热下油100ml，下牛肉末，煸炒出水气，肉要熟透，加郫县豆瓣酱100g，葱、姜末各20g，豆豉10g，翻炒一会儿，倒入水1L。锅开加料酒30ml，白糖10g，酱油15ml，倒入水淀粉勾芡，料汁即成。

❸ 将焯水后的豆腐滤水，料汁均匀浇在飞水后的豆腐上，放入万能蒸烤箱烹制入味，选择『单点分层煎烤』模式，时长5分钟即可。出锅后撒上香葱末，菜品即成。

WORKING PROCESS

1. Stir tofu with salt 5g, place into the baking pan, pour water to cover it, quick steam via the Universal Steam Oven, select "Single Point Stratified Steam" mode for 5 minutes. Turn the starch 30g into water-starch mash for later use.

2. Make the sauce at the same time. Pour oil 100ml into a heated wok, add minced beef and stir-fry till the beef is well cooked, add Pixian chili bean paste 100g, minced scallion 20g, minced ginger 20g, fermented soya beans 10g, stir-fry and pour water 1L, pour cooking wine 30ml after the water boiled, sugar 10g, soy sauce 15ml, thicken the soup with water-starch mash, the sauce is done.

3. Dress the sauce on the quick steamed tofu, then bake via the Universal Steam Oven, select "Single Point Stratified Bake" mode for 5 minutes, then sprinkle chopped scallion, the dish is done.

中国大锅菜

蒸烤箱卷（纪念版）

The Big-Wok-Made Cuisine of China, Food Volume of Steam Oven (Commemorative Edition)

菜品特点

特色 黑白豆腐，黑指的是猪血豆腐，白指的是北豆腐，因猪血为深红色，又名『红白豆腐』。烹制黑白豆腐，北方喜用酱烧法，川渝多香辣入味，制作本菜的大师为川菜名家，因而本菜谱的黑白豆腐为川府风味。

品味 这道菜香辣油润，口味厚重，豆腐软嫩，是下饭佐食的佳品。对其口味产生重要影响的除了郫县豆瓣酱，还有牛肉末。牛肉受热油烹时，不但牛肉本身有香味，而且还有一股浓郁的醇香味随之逸出，给人以醇厚浓郁之感，猪肉也有香味，但缺少醇香味，产生的口味过于单调。

营养价值 猪血富含维生素 B_2、维生素C、蛋白质、铁、磷等营养成分。猪血中的血浆蛋白被人体内的胃酸分解后，产生一种解毒、清肠分解物，能够与侵入人体内的粉尘、有害金属微粒发生化学反应，易于毒素排出体外。

豆腐营养丰富，含有铁、钙、磷、镁等人体必需的多种微量元素，还含有糖类、植物油和丰富的优质蛋白，素有『植物肉』之美称。

品相 此菜色泽红亮，红白相间，菜色饱含香辣之感，令人食欲顿生，为之垂涎。

魔芋烧鸭

Name: Braised Duck with Konjak

制作人：李加双　　中国烹饪大师

Made by: Jiashuang Li　　A Great Master of Chinese Cuisine

主 料 Main Ingredient
整 鸭：1750g 切块
Duck　1750g　Cut

配 料 Burdening
魔 芋：1000g 切条
Konjak　1000g　Stripped
青 笋：500g 切条
Green Bamboo Shoot　500g
Stripped
青 蒜：150g 切段
Garlic Sprout　150g　Cut

调 料 Seasoning
清 油 Oil.......................100ml
盐 Salt 15g

酱 油 Soy Sauce 20ml
郫县豆瓣酱 Pixian Chili Bean
　　　　　Paste...............100g
料 酒 Cooking Wine 45ml
啤 酒 Beer.....................300ml
白 糖 Sugar......................20g
姜 末 Minced Ginger 20g
葱 末 Minced Scallion.........20g

制作方法

❶ 鸭块入盆，加盐 5g，料酒 25ml，搅拌均匀，腌制 10 分钟。

❷ 烤盘刷油，将鸭块放入万能蒸烤箱飞水，选择『单点分层蒸煮』模式，时长 5 分钟。

❸ 烧开一锅水，加入盐 10g，将魔芋倒入锅中，水两开之后，可去除魔芋中的碱味。

❹ 锅热下油，倒入开水 500ml，加酱油 20ml，料酒 20ml，白糖 20g，啤酒 300ml，锅开即可。锅热下油 100ml，加郫县豆瓣酱 100g，将其煸透，加葱、姜末各 20g 炒香，

❺ 将魔芋和料汁倒入鸭块中，放入万能蒸烤箱炖熟，选择『单点分层煎烤』模式，时长 30 分钟，在出锅前 1 分钟加入青笋条和青蒜段，菜品即成。

WORKING PROCESS

1. Marinate the duck pieces in a basin, add salt 5g, cooking wine 25ml, stir evenly and marinate for 10 minutes.

2. Brush oil on the baking pan, and quick steam duck pieces by using the Universal Steam Oven, select "Single Point Stratified Steam" mode for 5 minutes.

3. Boil a wok of boiled water and add salt 10g, put the konjak in the wok, heat it till 2 times boiling.

4. Make the sauce. Pour oil 100ml into a heated wok, add Pixian chili bean paste 100g, stir-fry and add minced scallion 20g, minced ginger 20g, pour boiled water 500ml, add soy sauce 20ml, cooking wine 20ml, sugar 20g, beer 300ml, till boiled again.

5. Mix the konjak and the sauce with duck pieces, braise by using the Universal Steam Oven, select "Single Point Stratified Grill" mode for 30 minutes. Add green bamboo shoots strips and garlic sprouts strips at the last minute. The dish is done.

中国大锅菜

菜品名称·魔芋烧鸭
Name: Braised Duck with Konjak

菜 品 特 点

特色 魔芋本身有毒，需先以石灰水浸煮后方可制成食材，所以在烹调中要先用开水煮一下，去除碱味。魔芋主要产于我国四川地区，魔芋烧鸭也是川菜的一道经典菜式，数百年来经久不衰。今天，许多减肥人士更加喜爱魔芋，它可以清肠、降脂，而且因为其特性容易让人产生饱腹感，但热量却又很低。

品味 鸭肉肥酥，滋味咸中带鲜，辣而有香，啤酒使鸭肉的味道更加浓厚，还透着一股啤酒的清香。魔芋酥软细腻，充分吸收了料汁中的味道和鸭肉浓郁的香味，其美味不亚于鸭肉。青笋、青椒与蒜苗又能起到调解鸭肉油腻的作用，又丰富了营养价值。

品相 这道菜色泽红亮，给人以充分的香辣之感，添加几种蔬菜，则红绿相间，菜色诱人。

营养价值 魔芋含有大量甘露糖苷、维生素、植物纤维及一定量的黏液蛋白，具有奇特的保健作用和医疗效果，被人们誉为「魔力食品」，可活血化瘀，解毒消肿，宽肠通便，是健康饮食的不二之选。鸭肉中含有丰富的蛋白质，容易被人体吸收，所含B族维生素和维生素E较其他肉类多，能有效抵抗神经炎和多种炎症，广受百姓喜爱。

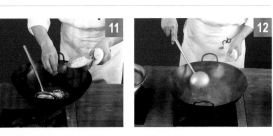

虾肉蒸蛋

Name: Stuffed Balsam Pear

制作人：李加双　　中国烹饪大师

Made by: Jiashuang Li　　A Great Master of Chinese Cuisine

主　料　Main Ingredient

鸡　蛋：1000g　打散
Egg　1000g　Stirred

虾　肉：250g　去虾线
Shrimp　250g　Cleaned

配　料　Burdening

香　菇：250g　切末
Mushroom　250g　Minced

青　豆：300g
Green Bean　300g

调　料　Seasoning

清　油　Oil 45ml

盐　Salt 7g

香　油　Sesame Oil 20ml

胡椒粉　Ground Pepper 3g

料　酒　Cooking Wine 35ml

淀　粉　Starch......................20g

姜　末　Minced Ginger 10g

葱　末　Minced Scallion.........10g

备　注　Tips

1. 如果不加料油就搅拌鸡蛋，容易产生太多气泡。

2. 将水蛋的烤盘用保鲜膜敷上，会使水蛋即使经过很长时间的蒸制也依然很鲜嫩。

1. Pour cooking wine before stirring the eggs to avoid to creating bubbles.

2. Covering the baking pan with a layer of plastic wrap to make the egg custard to keep tender and fresh.

中国大锅菜

菜品名称·虾肉蒸蛋
Name: Stuffed Balsam Pear

制作方法

❶ 将虾仁入小盆腌制，加盐2g，料酒5ml，干淀粉5g，抓匀，腌制15分钟。将15g淀粉制成水淀粉备用。

❷ 将鸡蛋打入盆中，先不要搅拌，加料酒15ml，胡椒粉3g，清油15ml。此时再搅拌鸡蛋，将其打散，加水。水和鸡蛋的比例是1：1即可。

❸ 将鸡蛋液倒入烤盘，敷上一层保鲜膜，要粘紧烤盘四周，尽量不要漏气，放入万能蒸烤箱蒸制，选择『单点分层蒸煮』模式，时长15分钟。

❹ 蒸蛋时炒制料汁，锅热下油30ml，加葱、姜末各10g爆香，放入青豆和香菇末，倒入水300ml，锅开后加盐5g，料酒15ml，倒入水淀粉勾芡，然后放入虾肉，烧制一会儿，待虾肉成熟，倒入香油20ml，料汁即成。

❺ 将料汁均匀浇在蒸好的水蛋上，菜品即成。

WORKING PROCESS

1. Marinate the shrimp in a basin. Add salt 2g, cooking wine 5ml, starch 5g, well stir and marinate for 15 minutes. then turn the starch 15g into water-starch mash for later use.

2. Break an egg into basin, pour cooking wine 15ml, ground pepper 3g, oil 15ml, stir the egg and pour water. The ratio of egg to water is 1 : 1.

3. Pour egg mash into baking pan, cover a layer of plastic wrap to the baking pan strictly cling around the pan and try not to leak. Steam by the Universal Steam Oven, select "Single Point Stratified Steam" mode for 15 minutes.

4. Make the sauce at the same time. Pour oil 30ml into a heated wok, add minced scallion 10g, minced ginger 10g, quick deep-fry it till fragrance. Then add green beans and minced mushroom, pour water 300ml, add salt 5g after water boiled, cooking wine 15ml, thicken the soup with water-starch mash and then put shrimp. Braise for a while till the shrimp is cooked, pour sesame oil 20ml and the sauce is done.

5. Dress the sauce on the steamed egg, the dish is done.

中国大锅菜

蒸烤箱卷（纪念版）

The Big-Wok-Made Cuisine of China, Food Volume of Steam Oven（Commemorative Edition）

菜品特点

特色 这是一道充满了儿时回忆的家常菜，许多人都盼望妈妈做这道菜，南方人喜欢称之为『蒸水蛋』，北方人则称为『鸡蛋羹』。

品味 虾仁味道十分鲜美，与水蛋一起蒸制，其鲜味与鸡蛋的香味融合，成为至鲜至香的一道菜品，且二者皆为质嫩之物，便于咀嚼，口感上佳。青豆的使用则给这道菜增添了绿色的气息，与软嫩的水蛋形成对比，在口感上有所调剂。

营养价值 虾肉营养价值丰富，性温，富含蛋白质，而脂肪含量较低，易于消化，是滋补的佳品。鸡蛋几乎含有人体必需的所有营养物质，如蛋白质、脂肪、卵黄素、卵磷脂、维生素和铁、钙、钾，被人们称作『理想的营养库』。

品相 水蛋呈金黄之色，上面铺满一层青红相间的青豆、虾仁，菜色具有富贵之气，表面一层薄薄的芡汁，将鲜味都锁在里面。

海米冬瓜

清蒸鱼块

菜品名称

宫保鸡丁

Name: Kung Pao Chicken

制作人：李建国　　中国烹饪大师

Made by: Jianguo Li　　A Great Master of Chinese Cuisine

主　料　Main Ingredient
鸡腿肉：2000g　切丁
Drumstick　2000g　Cut

配　料　Burdening
黄　瓜：1000g　切丁
Cucumber　1000g　Cut
烤干花生米：500g　去皮
Baked Peanut　500g　Peeled

调　料　Seasoning
油 Oil 100ml
盐 Salt 50g
醋 Vinegar...........................80ml

白　糖　Sugar......................180g
酱　油　Soy Sauce 50ml
老　抽　Dark Soy Sauce......50ml
辣椒段　Chili 80g
淀　粉　Starch......................50g
花　椒　Chinese Prickly Ash 8g
料　酒　Cooking Wine 100ml
姜　末　Minced Ginger 40g
葱　节　Scallion...................100g

备　注　Tips
1. 此菜腌制蒸好的鸡丁和炒料汁是两个十分重要的调味阶段，目的是把菜的底味做足，并起到调节颜色作用，并且不同地区可以根据自己口味调节咸度。
2. 黄瓜丁最后铺在鸡丁上，不易变黄，更加鲜亮明丽。

1. The marinating and seasoning are 2 important steps of cooking the dish. The purpose of doing that is to ensure the taste delicious enough, and make color as well as possible.
2. The last step is paving the cucumber pieces on the chicken which can keep the color of cucumber.

中国大锅菜

蒸烤箱卷（纪念版）

The Big-Wok-Made Cuisine of China, Food Volume of Steam Oven（Commemoriavte Edition）

制作方法

① 将鸡丁放入万能蒸烤箱进行滑油，烤盘刷底油，选择『蒸制蔬菜-99℃』模式，时长3分钟，鸡丁蒸好后滗掉底盘的油和水。再将黄瓜放入烤盘，选择『蒸制蔬菜-99℃』模式，蒸1分钟取出。

② 将滑油后的鸡丁放入盆中进行腌制，依次放入盐30g，料酒50ml，酱油20ml，老抽50ml。拌匀后放入烤盘，抹平，选择『单点分层炙烤』模式，时长3分钟。

③ 在烤制的过程中制作料汁，烧热炒锅下油100ml，放入辣椒段80g，炒香后滗出辣椒备用，放入花椒8g，炒香后捞出，放入葱节100g，姜末40g，酱油30ml，料酒50ml，白糖80g，盐20g，醋80ml，水350ml，烧开后加水淀粉50g进行勾芡，芡略薄一些。

④ 将料汁倒入烤好的鸡丁中拌匀，盛入布菲芯盒中，上面铺上蒸制好的黄瓜丁，最后再铺上烤干的香酥花生米和炸熟的辣椒段，一道香气四溢、色泽靓丽的宫保鸡丁便制作完成。

WORKING PROCESS

1. Quick stir-fry the chicken by using the Universal Steam Oven, brush oil at the baking pan, select "Steam Vegetable-99℃" mode for 3 minutes, filter the oil and water in the chicken. Then place the cucumber into the baking pan, select "Steam Vegetable-99℃" mode for 1 minute.

2. Marinate the chicken in a basin, then add salt 30g, cooking wine 50ml, soy sauce 20ml, dark soy sauce 50ml, stir well and then place into baking pan, select "Single Point Stratified Grill" mode for 3 minutes.

3. Make the sauce at the same time. Pour oil 100ml into a heated wok, then add chili 80g, stir-fry till fragrance and set aside for later use. Then add Chinese prickly ash 8g, stir-fry till fragrance and get out of the wok. Then add scallion 100g, minced ginger 40g, soy sauce 30ml, cooking wine 50ml, sugar 80g, salt 20g, vinegar 80ml, water 350ml, boil it and thicken the soup with starchy mash 50g.

4. Pour the sauce on the chicken, stir well. Then place it into buffet pot. Pave the steamed cucumber pieces, then pave a layer of baked peanuts and fried chili. The dish is done.

中国大锅菜

菜品名称·宫保鸡丁

Name: Kung Pao Chicken

菜品特点

特色 这是一道赫赫有名的川菜，起源于清朝末年。『宫保』指的是四川总督丁宝桢，此菜因此而得名。他有个荣誉官职叫『太子少保』，人们称呼获此职位的人为『宫保』，

丁宝桢喜食辣椒炒制的鸡肉和花生米，这道私房菜后来就流传开来，成为火遍全国各地的一道菜肴，也是人们认识川菜的一张名片。宫保鸡丁在山东地区也十分流行，可能和丁宝桢长期主政山东地区有关系，他广受爱戴，政绩斐然，亦对山东的饮食产生了影响。

品味 宫保鸡丁由于太过出名，已经形成了『宫保』系的口味，而更精确地说，它属于『糊辣荔枝味』。糊辣是通过一定温度的油，将花椒和辣椒脂溶脱水焦化，产生一种奇妙香气，从而达到微辣不燥的效果；而荔枝味强调破口酸、回口甜，口感顺序是酸味大于甜味，酸甜味又大于咸味，从而达到酸甜中求咸鲜的目的，宫保鸡丁入口先酸后甜，又因为有了花椒、辣椒的加入，它还带上了一股糊辣味，因此就形成了川菜中独一无二的『糊辣荔枝味』。

品相 辣椒火红，鸡肉黄嫩，辅以翠绿的黄瓜丁，菜品颜色更加鲜艳。芡汁浓稠，挂在鸡丁上，酸甜之感令人口中生津。

营养价值 鸡肉肉质细嫩，滋味鲜美，并富有营养，有滋补养身的作用。鸡肉中蛋白质的含量比例很高，而且消化率高，很容易被人体吸收利用，有增强体力、强壮身体的作用。中医认为鸡肉性平、味甘，入脾、胃经，可益气、补精、添髓。

菜品名称

菊 花 鱼

Name: Chrysanthemum Shaped Fish Fillets

制作人：李建国　　中国烹饪大师

Made by: Jianguo Li　　A Great Master of Chinese Cuisine

主 料 Main Ingredient	盐 Salt 20g	淀 粉 Starch.....................200g
草 鱼：1000g 切片	料 酒 Cooking Wine 15ml	葱 段 Scallion 20g
Grass Carp　1000g　Sliced	胡椒粉 Ground Pepper 3g	姜 片 Sliced Ginger 20g
	番茄酱 Ketchup..................300g	
调 料 Seasoning	白 醋 White Vinegar........100ml	
清 油 Oil......................100ml	白 糖 Sugar.....................100g	

中国大锅菜

菜品名称·菊花鱼

Name: Chrysanthemum Shaped Fish Fillets

制作方法

❶ 首先要将鱼肉用刀工加以修饰，先片鱼片，片出尽量薄的鱼片，保持鱼皮相连，片出5片鱼片后将鱼皮切断，把鱼皮相连的5片鱼片整齐地放在案板上，用刀尖垂直于鱼皮的方向切出梳子刀。

❷ 将鱼肉入盆腌制，加葱、姜各20g，料酒15ml，盐5g，胡椒粉3g，用手轻轻拌匀，不要用力，以防鱼肉破碎，腌制10分钟。

❸ 将腌制好的鱼肉沾满干淀粉，抖掉淀粉较厚的部分，将每一『朵』菊花鱼用牙签固定在黄瓜结上，放入万能蒸烤箱炸制。手动控制烤箱模式，湿度为10%，温度260℃，时长4分钟，然后取出淋油，再用240℃温度、4挡风速，复烤4分钟，即可形成菊花造型。

❹ 烧制料汁，锅中倒入水500ml烧开，加入300g番茄酱，盐15g，白醋100ml，白糖100g，锅开后倒入200g淀粉制成的水淀粉勾芡，浇上100ml热油，料汁即成。

❺ 将料汁浇在菊花鱼上，菜品即成。

WORKING PROCESS

1. First of all, cut and slice the fish fillets as thin as possible, keep the skin connected, cut the fish every 5 pieces, then place them on cutting board, cut the fish skin vertically as a comb.

2. Marinate the fish in a basin. Add scallion 20g, ginger 20g, cooking wine 15ml, salt 5g, ground pepper 3g, knead it gently, marinate about 10 minutes.

3. Coat dry starch all over pickled fish, then shake off the extra starch, fasten every chrysanthemum-shaped fish to the cucumber knot with a toothpick, and put it in the universal steaming oven. Control the oven mode manually with humidity at 10%, temperature at 260℃ for 4 minutes, then take it out, leach oil on the fish, and then bake it at 240℃ and 4-gear wind speed for 4 minutes to form the shape.

4. Make the sauce. Pour water 500ml into a wok and boiled, add ketchup 300g, salt 15g, white vinegar 100ml, sugar 100g, thicken the soup while it boils then pour hot oil 100ml, the sauce is done.

5. Pour the sauce on the fish fillets and dish is done.

菜品特点

特色 这是一道造型美观的经典菜肴，其酸甜口感尤其受儿童喜爱。制作此菜，第一关就是刀工，将鱼肉经过刀工修饰，放入锅中炸出菊花的造型。传统烹饪中，都会使用油锅炸制，由于鱼肉飘在油中，可以形成立体感较强的菊花造型，而使用万能蒸烤箱制作此菜，堪称作者一大创意。为了保持菊花鱼的立体造型，他巧妙使用了黄瓜节做支架，将鱼肉架起来。烹制时，先烤制再淋油，并手动控制万能蒸烤箱内的温度、湿度和风速，制作出十分接近油炸效果的菊花鱼。

品味 鱼肉趁热浇上料汁，散发出甜甜的香气，经过炸制后，外酥里嫩。每一口下去，都能品尝到鱼肉的鲜美和料汁的酸甜。

品相 菊花鱼，顾名思义，以其酷似菊花的造型而得名，菜肴出品后，刀工的水平便一见高下，菊花的花瓣要均匀，整体造型圆润，有如花开。

营养价值 草鱼含有丰富的不饱和脂肪酸，对血液循环有利，是心血管病人的良好食物。它还含有丰富的硒元素，经常食用有抗衰老、养颜的功效，而且对肿瘤也有一定的防治作用。此外，对于身体瘦弱，食欲不振的人来说，草鱼肉嫩而不腻，可以开胃、滋补。

老醋烧肉

Name: Braised Pork with Old Vinegar

制作人：李建国　　中国烹饪大师

Made by: Jianguo Li　　A Great Master of Chinese Cuisine

主　料 Main Ingredient
五花肉：1750g　切块
Pork Belly　1750g　Cut

配　料 Burdening
土　豆：1000g　切块
Potato　1000g　Cut

胡萝卜：500g　滚刀块
Carrot　500g　Cut

调　料 Seasoning
盐 Salt 30g
米　醋 Rice Vinegar........ 1250ml
料　酒 Cooking Wine 120ml

酱　油 Soy Sauce 20ml
老　抽 Dark Soy Sauce 5ml
胡椒粉 Ground Pepper 5g
白　糖 Sugar 10g
葱　片 Sliced Scallion........... 20g
姜　片 Sliced Ginger 20g

制作方法

1 将五花肉块放入万能蒸烤箱飞水，选择『单点分层蒸煮』模式，时长3分钟。

2 将飞水后的肉块加老抽5ml腌制上色，再加葱片20g，姜片20g拌匀，放入万能蒸烤箱烤制，选择『单点分层炙烤』模式，时长20分钟。

3 调制生汁，盆内倒入温水700ml，加米醋1250ml，料酒120ml，酱油20ml，盐30g，胡椒粉5g，白糖10g，搅拌均匀。

4 将生汁倒入烤好的肉块中，加入土豆和胡萝卜，盖上盖，炖至成熟，选择『肉类—焖炖』模式，时长70分钟。

WORKING PROCESS

1. Quick steam the pork belly by using the Universal Steam Oven, select "Single Point Stratified Steam" mode for 3 minutes.

2. To color the steamed pork via dark soy sauce 5ml, add sliced scallion 20g, sliced ginger 20g, stir evenly, bake by using the Universal Steam Oven, select "Single Point Stratified Grill" mode for 20 minutes.

3. Make the dressing. Pour warm water 700ml into a basin, add rice vinegar 1250ml, cooking wine 120ml, soy sauce 20ml, salt 30g, ground pepper 5g, sugar 10g, stir evenly.

4. Pour the dressing on the pork, add carrot and potatoes, braise till cooked, select "Meat-Braise" mode for 70 minutes.

中国大锅菜

菜品名称·老醋烧肉

Name: Braised Pork with Old Vinegar

菜品特点

特色 老醋烧肉是一道成都公馆菜，流传不广，烹制难度较高，醋多易过酸，且容易产生一股『馊』味。这道菜美味的关键是醋量的把握，醋仅是起调味作用和『引发剂』的功效，用醋酸加速肉的纤维化速度，将脂肪消除干净，肉质会更加细嫩鲜美。

品味 食用这道菜，第一口品尝和之后的品尝，味觉的感受是不同的。初尝，味道十分清新，醋酸较冲，直刺味觉，极大地打开人的味蕾；之后的品尝，感受较多的是醋味的香醇，味感更加悠长，酸味不再那么重，而是无穷的回味。

品相 用老醋烧制，颜色较深。菜品一端出来，有点其貌不扬之感，但一闻到味道，所有的疑虑顿消。

营养价值 五花肉含有丰富的优质蛋白质和必需的脂肪酸，并提供血红素（有机铁）和促进铁吸收的半胱氨酸，能改善缺铁性贫血；具有补肾养血，滋阴润燥的功效。土豆含有优质蛋白质、碳水化合物、铁、维生素B和维生素C等物质。土豆的特殊黏蛋白，不但有润肠作用，还有促进脂类代谢的作用。

炒藕丁

红烧鲈鱼

锅塌豆腐

Name: Baked Tofu with Eggs

制作人：李智东　　中国烹饪大师

Made by: Zhidong Li　　A Great Master of Chinese Cuisine

主 料 Main Ingredient	**调 料** Seasoning	淀 粉 Starch......................50g
豆 腐：4000g 切片	清 油 Oil.........................100ml	白 糖 Sugar........................15g
Tofu　4000g　Sliced	盐 Salt 25g	姜 末 Minced Ginger 3g
	鸡 粉 Chicken Powder 5g	蒜 末 Minced Garlic..............3g
配 料 Burdening	味 精 Monosodium	葱 末 Minced Scallion.........15g
鸡 蛋：3 个 打散	Glutamate 18g	
Egg　Three　Stirred	胡椒粉 Ground Pepper 5g	

中国大锅菜

制作方法

❶ 豆腐倒入盆内，加盐5g、味精3g腌制15分钟。将2个鸡蛋打入碗内，加盐少许，淀粉20g拌匀，制成鸡蛋糊。将30g淀粉制成水淀粉备用。

❷ 将腌制好的豆腐放入万能蒸烤箱煎制，烤盘上刷底油，将豆腐片摆入，豆腐片上刷明油，选择「单点分层炙烤」模式，时长8分钟，煎制两面金黄即可。

❸ 然后烧制料汁，锅热下油80ml，油热下葱花15g，姜末、蒜末各3g，倒入水500ml，加盐15g、味精15g、糖15g、鸡粉5g、胡椒粉5g，锅开后倒入水淀粉勾成玻璃芡，料汁即成。

❹ 将料汁均匀浇在煎制好的豆腐上，用万能蒸烤箱再次烹制，选择「单点分层煎烤」模式，时长20分钟。

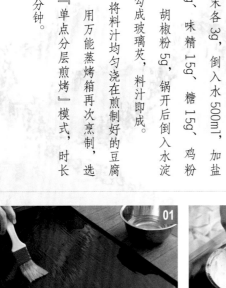

WORKING PROCESS

1. Pour the tofu in a basin. Add salt 5g, MSG 3g for 15 minutes. Put 2 eggs, add a few salt, starch 20g to stir evenly, turn it into egg-mash. Turn the starch 30g into water-starch mash for later use.

2. Bake the marinated tofu by using the Universal Steam Oven, brush the oil at the bottom of the baking pan as well as the surface of tofu, select "Single Point Stratified Grill" mode for 8 minutes till its both sides turn into golden yellow.

3. Make the sauce. Pour oil 80ml into a heated wok, add chopped scallion 15g, minced ginger 3g, minced garlic 3g, pour water 500ml, MSG 15g, chicken powder 5g, ground pepper 5g thicken the sauce after the water boiled, then the sauce is done.

4. Pour the sauce on the baked tofu, bake again by using the Universal Steam Oven. Select "Single Point Stratified Bake" mode for 20 minutes.

中国大锅菜

The Big-Wok-Made Cuisine of China, Food Volume of Steam Oven（Commemorative Edition）

蒸烤箱卷（纪念版）

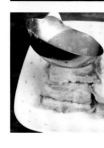

菜品特点

特色 锅塌豆腐是一道鲁菜名菜，源自山东地区，早在明代就有出现，清代乾隆年间荣升为宫廷菜，并传遍各地。锅塌是鲁菜独有的一种烹调方法，将原材料腌制后，裹上淀粉、蛋液进行煎制，然后加入料汁烧至收汁即可，用锅塌法还可以制作锅塌鱼、锅塌里脊等。

品味 豆腐起源于中国，在漫长的历史中，中国人积累了很多烹饪豆腐的方法，其中的关键是如何让味道附着其上，锅塌豆腐就是成功的典范。豆腐本身软嫩可口，裹上淀粉、蛋液后，又增添了香味，加入料汁微火塌制，十分入味。料汁为玻璃芡，咸鲜中略有一点微甜，与豆腐的软嫩相得益彰。

品相 蛋液经过煎制后呈现出灿烂的金黄色，玻璃芡汁挂在上面，晶莹剔透，每一块豆腐都像一件艺术品一样，光芒闪耀。

营养价值 这道菜清淡可口，是养生佳品。豆腐营养丰富，含有铁、钙、磷、镁等人体必需的多种微量元素，还含有糖类、植物油和丰富的优质蛋白，素有『植物肉』之美称，有高蛋白，低脂肪，降血压，降血脂，降胆固醇的功效。大豆蛋白属于完全蛋白质，其氨基酸组成比较好，人体所必需的氨基酸几乎都有，并且十分容易被人体消化、吸收。

胡椒虾

❧

Name: Baked Shrimp with Black Pepper

制作人：李智东　　中国烹饪大师

Made by: Zhidong Li　　A Great Master of Chinese Cuisine

主 料 Main Ingredient	花生仁：250g	味 精 Monosodium Glutamate..5g
白 虾：2500g 去虾线	Peanut 250g	胡椒粉 Ground Pepper 5g
Shrimp 2500g Cleaned		料 酒 Cooking Wine 30ml
	调 料 Seasoning	姜 片 Sliced Ginger 30g
配 料 Burdening	料 油 Spicing Oil 30ml	葱 段 Scallion 50g
香辣酥：1000g 切块	盐 Salt 15g	
Flavor-spicy Sauce 1000g Diced	淀 粉 Starch......................30g	

中国大锅菜

蒸烤箱卷（纪念版）

The Big-Wok-Made Cuisine of China, Food Volume of Steam Oven（Commemorative Edition）

制 作 方 法

❶ 首先将生虾进行腌制，虾倒入盆中，加入葱段 50g，姜片 30g，料酒 30ml，撒入盐 15g，胡椒粉 5g，味精 5g，淀粉 30g，料油 30ml，搅拌均匀，腌制 15 分钟。

❷ 烤盘内刷油，将腌制好的鲜虾放入万能蒸烤箱炸制，选择『鱼类－高温炙烤』模式，时长 5 分钟。

❸ 将香辣酥和花生仁倒入炸制好的虾中，即可盛盘。

WORKING PROCESS

1. Marinate the shrimps. Put them into a basin, add scallion 50g, sliced ginger 30g, cooking wine 30ml, sprinkle the salt 15g, ground pepper 5g, MSG 5g, starch 30g, spicing oil 30ml, stir evenly, marinate it for 15 minutes.

2. Brush the oil at the bottom of the baking pan, deep-fried the shrimp by using the Universal Steam Oven, select "Fish-High Temperature Grill" mode for 5 minutes.

3. Add the flavor-spicy sauce and peanut, the dish is done.

中国大锅菜

菜品名称·胡 椒 虾
Name: Baked Shrimp with Black Pepper

菜品特点

特色 这是来自祖国宝岛台湾的一道家常菜，台湾临海，盛产鱼虾，形成了多种多样的吃虾方法，胡椒虾就是其中一种。

品味 胡椒虾为炸制菜品，外皮酥脆，内里十分软嫩，肉质鲜美，经过胡椒粉调味，毫无腥气。

品相 虾烹制成熟后变成红色，而裹上的一层薄薄淀粉糊经过炸制后色泽金黄，使得这道菜充满富贵之气，十分鲜艳夺目。

营养价值 虾营养极为丰富，含蛋白质是鱼、蛋、奶的几倍到几十倍；还含有丰富的钾、碘、镁、磷等矿物质及维生素A、氨茶碱等成分，且其肉质和鱼一样松软，易消化，不失为营养佳品，对健康极有裨益；对身体虚弱以及病后需要调养的人也是极好的食物。

菜品名称

淮山烧牛腩

Name: Braised Sirloin with Chinese Yam

制作人：李智东　　中国烹饪大师

Made by: Zhidong Li　　A Great Master of Chinese Cuisine

主　料　Main Ingredient
牛 腩：2500g　切块
Sirloin　2500g　Diced

配　料　Burdening
淮山药：1500g　切块
Chinese Yam　1500g　Diced

调　料　Seasoning
清　油　Oil.......................100ml
盐　Salt 30g
老　抽　Dark Soy Sauce 10ml
味　精　Monosodium
　　　　Glutamate.................10g
蚝　油　Oyster Sauce............20g

料　酒　Cooking Wine 30ml
白　糖　Sugar........................20g
淀　粉　Starch.......................30g
葱　花　Minced Scallion.........30g

中国大锅菜

菜品名称·淮山烧牛腩

Name: Braised Sirloin with Chinese Yam

制作方法

❶ 首先将牛腩和山药进行飞水处理：摆入蒸盘中，放入万能蒸烤箱飞水，选择『单点分层蒸煮』模式，牛腩和山药都为5分钟。将30g淀粉制成水淀粉备用。

❷ 在飞水时炒汁，锅热下油100ml，油热下葱花30g，老抽10ml，料酒30ml，蚝油20g，倒入水3L，加盐30g，味精10g，糖20g，锅开倒入水淀粉勾芡，芡汁要稍薄些。

❸ 将飞水后的牛腩和山药倒在一起，浇上料汁拌匀，放入万能蒸烤箱炖制，选择『肉类—煎烤』模式，时长120分钟。

WORKING PROCESS

1. Steam the Chinese yam and sirloin at first by using the Universal Steam Oven, select "Single Point Stratified Steam" mode for 5 minutes. Turn the starch into water-starch mash for later use.

2. Make the sauce at the same time, pour oil 100ml into a heated wok, add chopped scallion 30g, dark soy sauce 10ml, cooking wine 30ml, oyster sauce 20g, water 3L, salt 30g, MSG 10g, sugar 20g, thicken the sauce by using the water-starch mash, but not too thick.

3. Mix the sirloin and Chinese yam together, pour the sauce, braised at the Universal Steam Oven, select "Meat-Bake" mode for 120 minutes.

中国大锅菜

蒸烤箱卷（纪念版）

The Big-Wok-Made Cuisine of China, Food Volume of Steam Oven (Commemorative Edition)

09

10

11

12

菜品特点

特色 这是一道经典家常菜，几乎会出现在每一个家庭的餐桌上。牛腩即牛腹部及靠近牛肋处的松软肌肉，是取自肋骨间的去骨条状肉，瘦肉较多，脂肪较少，筋也较少，适合红烧或炖汤。

品味 牛腩肉味浓郁，口感肥厚而醇香，肉质有韧性，咀嚼时可以感受到满口的香味；淮山药经过炖制，软烂入味，肉香之余，能感觉到山药本身的清香。

品相 本菜谱采用红烧做法，经过长达2个小时的炖制，牛肉和山药已经十分入味，视觉上，整个菜品呈现出酱香之色，使人胃口大开。

营养价值 山药是营养很丰富的食物，具有很高的药用价值，味甘、性平、入肺、脾、肾经，不燥不腻，具有健脾补肺、益胃补肾、固肾益精的功效。牛肉是高蛋白低脂肪的食材，营养价值十分高，氨基酸组成比猪肉更接近人体需要，能提高机体抗病能力，且有补中益气、滋养脾胃、强健筋骨的功效，并能起暖胃的作用，十分适合在冬天食用。

菜品名称

珍珠丸子

Name: Pearl-Shaped Meatballs

制作人：李智东　　中国烹饪大师

Made by: Zhidong Li　　A Great Master of Chinese Cuisine

主　料　Main Ingredient
五花肉：2500g　切末
Pork Belly　2500g　Minced

配　料　Burdening
糯　米：1250g　温水泡
Sticky Rice　1250g
Soaked in warm water
鸡　蛋：4 个
Egg　Four

调　料　Seasoning
香　油　Sesame Oil 20ml
盐　　Salt 30g
白　糖　Sugar.......................15g
淀　粉　Starch.......................30g
十三香　13-flavor 5g
鸡　粉　Chicken Powder 5g
料　酒　Cooking Wine 20ml
姜　末　Minced Ginger 5g
蒜　末　Minced Garlic..............5g
葱　末　Minced Scallion...........5g

备　注　Tips
1. 糯米一定要提前用温水泡制，否则易夹生。
2. 调制馅料不要放老抽，以保证丸子蒸出来洁白漂亮。
1. The sticky rice should be soaked in advance by warm water to make sure the sticky rice is well boiled.
2. Don't pour dark soy sauce during making the stuff to make sure the color of the meatballs are white.

中国大锅菜

The Big-Wok-Made Cuisine of China, Food Volume of Steam Oven（Commemorative Edition）

蒸烤箱卷（纪念版）

制作方法

❶ 将糯米用温水泡制 30 分钟，控干水分备用。

❷ 腌制五花肉馅，肉馅入盆，加料酒 20ml，盐 30g，白糖 15g，十三香 5g，鸡粉 5g，葱、姜、蒜末各 5g，打入 4 个鸡蛋，用手搅拌，摔打上劲，上劲后加淀粉 30g，香油 20ml，拌匀后腌制 15 分钟。

❸ 将五花肉馅做成 1 两左右的丸子，滚上一层糯米，摆入烤盘蒸制，选择『肉类—单点分层蒸煮』模式，时长 10 分钟。

WORKING PROCESS

1. Soak the sticky rice in warm water for 30 minutes. Take the rice out till dry.

2. Make the pork belly meat stuff, put the stuff into a basin, add cooking wine 20ml, salt 15g, sugar 15g, 13-flavor 5g, chicken powder 5g, minced scallion 5g, minced ginger 5g, minced garlic 5g, add 4 eggs and well stir. Then add starch 30g, sesame oil 20ml, stir evenly and marinate 15 minutes.

3. Make the meatballs (each ball is about 50g), coated with sticky rice, put them into baking pan to steam, select "Meat-Single Point Stratified Steam" mode for 10 minutes.

中国大锅菜

菜品名称·珍珠丸子
Name: Pearl-Shaped Meatballs

菜品特点

特色 这是一道湖北的地方特色菜，相传起源于沔阳（今湖北省仙桃市）地区，沔阳为鱼米之乡，物产丰富，其地蒸菜流行，有『沔阳三蒸』之传，这道珍珠丸子就是其中一道招牌菜。

品味 这道菜清香细嫩，鲜香适口。五花肉为猪肉中的上品，肉香浓郁，经过摔打上劲后滑弹适口，表面一层糯米则充分吸收了肉香，入口软糯，略有粘牙，将猪肉的香味留存于口中。

品相 本菜为蒸制而成，保持了食材的原貌，糯米白中微透猪肉的红色，每一粒米都颗粒分明，晶莹剔透。

营养价值 糯米营养价值丰富，含有蛋白质、脂肪、糖类、钙、磷、铁、维生素 B_1、维生素 B_2、烟酸等物质。中医认为它是温补强壮食品，具有补中益气，健脾养胃，止虚汗之功效，对食欲不佳，腹胀腹泻有一定缓解作用。五花肉性平，味甘咸，含有丰富的蛋白质及脂肪、碳水化合物、钙、磷、铁等成分。猪肉是日常生活的主要副食品，具有补虚强身，滋阴润燥、丰肌泽肤的作用。

五彩冬瓜

西芹羊肉

菜品名称

芝士烤鱼

Name: Grilled Long Li Fish with Cheese

制作人：林 进　　中国烹饪大师

Made by: Jin Lin　　A Great Master of Chinese Cuisine

主 料 Main Ingredient

鱼 肉：1000g 切片

Long Li Fish　1000g　Sliced

调 料 Seasoning

盐 Salt 16g

胡椒粉 Ground Pepper 8g

黄 油 Butter 160g

鸡 蛋 Egg ....2 个 (pcs)

面 粉 Powder 60g

牛 奶 Milk.....................400ml

芝士粉 Cheese Powder 15g

中国大锅菜

菜品名称·芝士烤鱼
Name: Grilled Long Li Fish with Cheese

制作方法

❶ 首先将鱼片放入盆中，加入盐1勺，打散蛋液，倒入鱼片中，搅拌均匀，让鱼肉和蛋液充分结合，腌制15分钟。

❷ 烤盘刷底油，将腌制好的鱼片摆入盘中，放入万能蒸烤箱烤制，选择『单点分层炙烤』模式，时长5分钟。

❸ 烤鱼时炒制料汁，锅内加黄油130g，黄油化开后加面粉60g，用小火将面粉炒熟、炒透，倒入牛奶400ml，加温水300ml，继续炒制，锅开后加盐10g，胡椒粉5g，料汁即成。

❹ 将料汁用勺子均匀浇在烤好的鱼片上，撒上芝士粉15g，放入几个黄油粒约30g，放入万能蒸烤箱烹制，选择『单点分层炙烤』模式，时长3分钟，菜品即成。

（原文顶部说明：加盐5g，胡椒粉3g，拌匀，将2个鸡蛋打入碗）

WORKING PROCESS

1. Put the sliced fish into a basin. Add salt 5g, ground pepper 3g, stirred evenly and then break 2 eggs into a bow. Then add salt 1g, stir the eggs mash and pour it into the fish. Marinate the fish with egg mash for 15 minutes.

2. Brush oil at the bottom of the baking pan, place the sliced fish into the baking pan, grill by using the Universal Steam Oven, select "Single Point Stratified Grill" mode for 5 minutes.

3. Make the sauce at the same time. Put butter 130g into wok, add powder 60g after the butter melted, heat and stir-fry the powder with light fire, pour milk 400ml, pour warm water 300ml, continue to stir-fry till the water boiled. Then add salt 10g, ground pepper 5g, the sauce is done.

4. Pour the sauce evenly on the sliced fish, sprinkle the cheese powder 15g, add butter about 30g, then use the Universal Steam Oven, select "Single Point Stratified Grill" mode for 3 minutes, the dish is done.

中国大锅菜

蒸烤箱卷（纪念版）

The Big-Wok-Made Cuisine of China, Food Volume of Steam Oven（Commemorative Edition）

菜品特点

特色 我国的国宴有『堂宴』和『台宴』之分，分别指举办国宴最多的两个地方：人民大会堂和钓鱼台国宾馆，芝士烤鱼就是『台宴』的经典菜之一，驰名中外。芝士就是奶酪，由英文『cheese』音译而来，这道菜是由俄式大餐演变而来，采用西餐方法烹调而成。

品味 未及品尝，远远便能闻到奶酪浓郁的香味，黄油使味道增加了厚重感，鱼肉的软嫩和鲜香，美味异常。品尝之时不禁令人由衷感觉这道菜搭配合理，牛肉、羊肉或许味道太重，又少鲜美之味，而鱼肉无论从肉质还是味道方面，都和奶酪是完美搭配，二者味道不相克，而是完美融合在一起。

品相 这道菜的主色调就是金黄色，充满富贵之气，与国宴的档次十分相符。烤制时，表面的奶酪化掉，让人真是迫不及待想要品尝。

营养价值 鱼肉的选用可以多种多样，但一定要少刺，以鲜味更浓的海鱼为佳，本菜谱选用的是龙俐鱼，可谓恰到好处。龙俐鱼具有海产鱼类在营养上显著的优点，含有较高的不饱和脂肪酸，蛋白质容易消化吸收。其肌肉细嫩，口感爽滑，鱼肉久煮不老，无腥味和异味，属于高蛋白、低脂肪、富含维生素的鱼类。

菜品名称

肉丸酿鲜蛋

Name: Brewed Meat Balls with Quail Eggs

制作人：林　进　　中国烹饪大师

Made by: Jin Lin　　A Great Master of Chinese Cuisine

主 料 Main Ingredient
五花肉：750g　绞馅
Pork Belly　750g　Minced

配 料 Burdening
荸 荠：500g　去皮切碎
Water Chestnut　500g

Skinning and Chopped
鹌鹑蛋：100g
Quail eggs　100g

调 料 Seasoning
盐 Salt 10g
鸡 汤 Chicken Soup 1700ml

胡椒粉 Ground Pepper 8g
料 酒 Cooking Wine 10ml
淀 粉 Starch 5g
香 菜 Caraway 20g
葱 段 Scallion 60g
姜 片 Sliced Ginger 30g

制作方法

❶ 首先调制肉馅，五花肉绞馅和荸荠切碎入盆，加盐10g，胡椒粉5g，料酒10ml，打入一个蛋清，搅拌均匀，再加入淀粉5g拌匀，肉馅成团后用力摔打上劲。

❷ 将肉馅团成80g左右重量的肉丸，摆入盘中，在肉丸上面摁出一个窝，每个窝打入一个鹌鹑蛋，放入万能蒸烤箱蒸制定型，选择『单点分层蒸煮』模式，时长3分钟。

❸ 调制汤汁，锅内倒入鸡汤1.7L，加胡椒粉3g烧开。肉丸定型后，将鸡汤倒入盘中，加上葱段60g，姜片30g，放入万能蒸烤箱炖制，选择『单点分层蒸煮』模式，时长15分钟，出锅后撒上香菜，菜品即成。

WORKING PROCESS

1. Make the meat stuffing firstly. Minced the pork belly and water chestnut into a basin. Add salt 10g, ground pepper 5g, cooking wine 10ml. Add one egg white, stirred it evenly. Then 5g more of starch to stir well. Make the meat stuffing for later use.

2. Make the meat stuffing into meat balls, about 80g per each. Place them into the baking pan, pit at each meat ball. Put a quail egg into each pit. Steam by using the Universal Steam Oven, select "Single Point Stratified Steam" mode for 3 minutes.

3. Make the dressing, pour chicken soup 1.7L into the wok, add ground pepper 3g and boil it. Pour the chicken soup into the baking pan after the meat ball shaped, add scallion 60g, sliced ginger 30g, boil it by using the Universal Steam Oven, select "Single Point Stratified Steam" mode for 15 minutes. Then sprinkle caraway, the dish is done.

 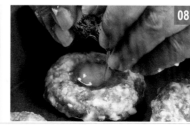 

中国大锅菜

菜品名称·肉丸酿鲜蛋

Name: Brewed Meat Balls with Quail Eggs

菜 品 特 点

特色 这道菜或许是许多人记忆中的『妈妈菜』，小时候正长身体，对营养的需求量很大，妈妈就会将鹌鹑蛋或鸡蛋打入肉丸中，肉和鸡蛋一起吃，可谓是大饱口福。

品味 拌肉馅时打入蛋清，更能锁住肉质中的水分，使肉质更加鲜嫩，鸡蛋亦是很多人的最爱，二者是最常见的两种荤味，其浓郁的香气相互融合，将口味又提升到一个新的层次。

品相 这是一道视觉上十分可爱的菜品，有种观点认为，黄色最有助于缓解人的情绪，看着肉丸上面鲜黄的蛋黄，作为食客，也许会心情大好，不知是否又会想到『少年不识愁滋味』呢？

营养价值 这是一道滋补佳肴，最适合儿童或需要补充营养之人，可谓美味与营养兼得。鹌鹑蛋几乎含有人体必需的所有营养物质，如蛋白质、脂肪、卵黄素、卵磷脂、维生素和铁、钙、钾，被人们称作『理想的营养库』。猪肉含有丰富的优质蛋白质和必需的脂肪酸，并提供血红素（有机铁）和促进铁吸收的半胱氨酸，能改善缺铁性贫血，具有补肾养血，滋阴润燥的功效。

咸蛋黄烧豆腐

Name: Stewed Tofu with Salted Duck Egg Yolk

制作人：林 进　中国烹饪大师

Made by: Jin Lin　A Great Master of Chinese Cuisine

主 料　Main Ingredient
豆 腐：2000g 切丁
Tofu　2000g　Pieced

配 料　Burdening
咸鸭蛋：500g 切丁
Salted Duck Egg　500g　Pieced
松花蛋：300g 切丁
Preserved Egg　300g　Pieced

调 料　Seasoning
清 油　Oil...........................50ml
料 酒　Cooking Wine 10ml
酱 油　Soy Sauce 10ml
盐　Salt 10g
胡椒粉　Ground Pepper 5g
淀 粉　Starch......................20g
葱 片　Sliced Scallion...........15g
姜 片　Sliced Ginger 15g

蒜 片　Sliced Garlic..............50g

备 注　Tips
不同的松花蛋和咸鸭蛋蒸熟时间不同，蒸至熟透即可。
The steam time of preserved duck egg and salted duck egg is different, which is steamed till well-done.

中国大锅菜

菜品名称·咸蛋黄烧豆腐

Name: Stewed Tofu with Salted Duck Egg Yolk

制作方法

① 将松花蛋和咸鸭蛋放入万能蒸烤箱蒸至熟透。将豆腐放入万能蒸烤箱飞水，选择『单点分层蒸煮』模式，时长15分钟。将20g淀粉制成水淀粉备用。

② 豆腐焯水时烧制料汁，锅热下油50ml，加蒜片50g炒香，倒入松花蛋丁和咸鸭蛋清丁，加葱、姜片各15g，倒入料酒10ml，酱油10ml，不断翻炒，倒入水500ml，锅开后加盐10g，胡椒粉5g，倒入水淀粉勾芡，料汁即成。

③ 将料汁倒入飞水后的豆腐中，搅拌均匀，表面撒上咸蛋黄，放入万能蒸烤箱烹制入味，选择『单点分层蒸煮』，时长5分钟，菜品即成。

WORKING PROCESS

1. Steam the preserved eggs and salted duck eggs. Quick steam the tofu by using the Universal Steam Oven, select "Single Point Stratified Steam" mode for 15 minutes. Make the starch 20g into water-starch mash for later use.

2. Make the sauce at the same time. Pour oil 50ml into a heated wok, stir-fry with sliced garlic. Then add pieced preserved eggs and salted duck eggs, add sliced scallion 15g, sliced ginger 15g, pour cooking wine 10ml, soy sauce 10ml, keep stir-frying, pour water 500ml. Add salt 10g and ground pepper 10g after the water boiled, add the water-starch mash to thicken the sauce, then the sauce is done.

3. Pour the sauce into quick steamed tofu, well stirred, sprinkle the salt egg yolk, steam by using the Universal Steam Oven, select "Single Point Stratified Steam" mode for 5 minutes, the dish is done.

中国大锅菜

The Big-Wok-Made Cuisine of China, Food Volume of Steam Oven (Commemorative Edition)

蒸烤箱卷（纪念版）

菜品特点

特色 『咸鸭蛋』是一种中国特色的食品，是古人智慧的结晶，过去储存食物是摆在古人面前的一大难题，因而产生了很多腌制食品，咸鸭蛋就是其中之一。鸭蛋的鲜香经过腌制，便可以长时间保存下来，甚至已经远远超越其原有的味道。咸鸭蛋的美味在烹饪中得到了完美的释放，形成了很多相关菜品，烧豆腐就是其中一道经典菜。

品味 豆腐本身略带淡淡的香味，需要赋予其很浓的味道才好，咸蛋黄的鲜香在这道菜中得到了极大的释放，溶入汤汁中，每一块豆腐都能带来味蕾上的享受。这道菜蛋清亦不可少，清代随园食单中便谈到蛋黄蛋清二者不可分离：『黄白兼用，不可存黄去白，使味不全。』

品相 相信每一个吃过咸鸭蛋的人都无法拒绝蛋黄的味道，看着表面细碎的蛋黄块和淡黄色的汤汁，未及品尝，那种味道已隐约出现在口中。让人见其菜色，心已神往。

营养价值 咸鸭蛋营养丰富，不仅仅含有很多蛋白质、维生素，而且还有很多我们身体所需的矿物质和微量元素。豆腐营养丰富，不但含有铁、钙、磷、镁等人体必需的多种微量元素，还含有糖类、植物油和丰富的优质蛋白，素有『植物肉』之美称，有降血压、降血脂、降胆固醇的功效。

菜品名称

香草烤鸡腿

Name: Roasted Drumstick Whith Herb

制作人：林 进　　中国烹饪大师

Made by: Jin Lin　　A Great Master of Chinese Cuisine

主 料 Main Ingredient
鸡 腿：2000g 去骨
Drumstick　2000g　Boning

配 料 Burdening
胡萝卜：100g 切丝
Carrot　100g　Shred

洋 葱：100g 切丝
Onion　100g　Shred
芹 菜：100g 切条
Celery　100g　Cut

调 料 Seasoning
百里香 Thyme 10g

迷迭香 Rosemary 10g
干百里香 Dried Thyme............1g
干迷迭香 Dried Rosemary.......1g
黑胡椒碎 Ground Black Pepper..5g
盐 Salt 20g
红 酒 Red Wine................15ml
清 油 Oil..........................30ml
蒜 片 Sliced Garlic..............60g

制作方法

❶ 将百里香10g切成段、迷迭香10g摘下叶子备用。

❷ 炒制腌制鸡腿的料油，锅热下油30ml，加蒜片60g，将百里香和迷迭香倒入，炒出香味即可。

❸ 腌制鸡腿，将鸡腿肉入盆，加胡萝卜丝、洋葱丝、芹菜条各100g，撒入盐20g，黑胡椒碎5g，干迷迭香、干百里香各1g，倒入红酒15ml，用力抓揉。挤出蔬菜中的汁液，搅拌均匀，使配料与鸡肉充分结合。再倒入之前炒好的料油，拌匀后腌制30分钟。

❹ 将腌料倒入烤盘，鸡腿皮朝上摆入盘中，放入万能蒸烤箱烤制成熟，选择『单点分层炙烤，4号色』，时长10分钟，出锅后将蔬菜去掉，只留下鸡腿，菜品即成。

WORKING PROCESS

1. Cut the thyme 10g. Rosemary leaves 10g for later use.

2. Make the seasoning. Pour oil 30ml into a heated wok, add sliced garlic 60g, stir-fry the thyme and the rosemary till fragrance.

3. Marinate the drumstick in a basin. Add shred carrot 100g, shred onion 100g, celery 100g. Sprinkle salt 20g, ground pepper 5g, dried thyme 1g, dried rosemary 1g, pour red wine 15ml, knead it and stir evenly. Marinate the drumstick with seasoning that made just now for 30 minutes.

4. Place all the ingredients into baking pan. Roast by using the Universal Steam Oven, select "Single Point Stratified Grill, Color No. 4" for 10 minutes. Then get rid of the vegetables, the dish is done.

中国大锅菜

菜品名称·香草烤鸡腿

Name: Roasted Drumstick Whith Herb

菜品特点

特色 这是一道由西餐演化而来的菜品，从调料到烹饪方法都充满了「洋」气。香草选用西餐中常见的百里香和迷迭香，干、鲜两种皆用，两种香草皆为多年生草本，在西方家庭的庭院中亦有广泛种植。

品味 百里香芳香袭人，带有优雅、浓郁的麝香香味，它能把食物中的不同味道融在一起，这是因为它的味道温和又不太刺激之故。迷迭香气味很冲，有青草一样的清凉气味和甜樟脑的气息，烹调中使用的量不宜过大。红酒、洋葱、芹菜则使鸡肉软嫩、香甜，兼有去除腥味的作用。

品相 鸡腿外皮略焦，点缀几点干香草，散发诱人的芳香，切开后鸡肉红嫩，饱含汁水，软嫩适口。

营养价值 鸡肉肉质细嫩，滋味鲜美，由于其味较淡，因此可用于各种料理中。蛋白质的含量颇多，在肉类中，可以说是蛋白质最高的肉类之一，是属于高蛋白、低脂肪的食品。钾硫酸氨基酸的含量颇多，因此可弥补牛肉及猪肉的不足，同时鸡肉比其他肉类的维生素A含量多。

菜品名称

冬菜烧鸭块

Name: Braised Duck with Tianjin Preserved Cabbage

制作人：林 勇　　中国烹饪大师

Made by: Yong Lin　　A Great Master of Chinese Cuisine

主 料 Main Ingredient
鸭 肉：1500g 切块
Duck 1500g Pieced

配 料 Burdening
冬 菜：300g 切末
Tianjin Preserved Cabbage 300g
Minced

土 豆：1000g 切块
Potato 1000g Pieces
胡萝卜：500g 切块
Carrot 500g Pieced

调 料 Seasoning
清 油 Oil.........................100ml
盐 Salt 30g

酱 油 Soy Sauce 40ml
胡椒粉 Ground Pepper 5g
料 酒 Cooking Wine 30ml
白 糖 Sugar.......................10g
淀 粉 Starch......................50g
姜 Ginger............................70g
葱 Scallion 70g
料 油 Spicing Oil 10ml

中国大锅菜

菜品名称 · 冬菜烧鸭块
Name: Braised Duck with Tianjin Preserved Cabbage

制 作 方 法

❶ 将鸭块入盆腌制，加葱、姜片各50g，盐10g，料酒10ml，酱油10ml，胡椒粉3g，干淀粉50g，拌匀，腌制15分钟。

❷ 将土豆、胡萝卜入盆上底味，加盐10g，胡椒粉2g，拌匀，淋上料油10ml，继续拌匀。

❸ 将两个烤盘刷底油，分别倒入腌制好的鸭块和上好底味的辅料，放入万能蒸烤箱炸制，选择『单点分层炙烤』模式，时长6分钟。

❹ 炸制时炒制料汁，锅热下油100ml，油热加葱、姜末各20g，爆香。倒入冬菜翻炒，加料酒20ml，酱油30ml，倒入水2L，锅开后加盐10g，白糖10g，料汁即成。

❺ 将炸制好的主辅料倒入深烤盘，倒入料汁，放入万能蒸烤箱继续炖制，选择『单点分层煎烤』模式，时长20分钟，即可出锅。

WORKING PROCESS

1. Marinate duck pieces in a basin. Add sliced scallion 50g, sliced ginger 50g, salt 10g, cooking wine 10ml, soy sauce 10ml, ground pepper 3g, starch 50g, well stir and marinate for 15 minutes.

2. Marinate potato and carrot in a basin. Add salt 10g, ground pepper 2g, well stir and pour spicing oil 10ml.

3. Brush oil on the bottom of 2 baking pans. One baking pan for the duck and one for the cabbage. Deep-fry by using the Universal Steam Oven, select "Single Point Stratified Grill" mode for 6 minutes.

4. Make the sauce at the same time. Pour oil 100ml into a heated wok, then add minced scallion 20g, minced ginger 20g, deep-fry till fragrance, then add Tianjin preserved cabbage, pour cooking wine 20ml, soy sauce 30ml, water 2L, add salt 10g after water boiled, sugar 10g, the sauce is done.

5. Place all the ingredients into a deep baking pan, dress the sauce, braise by using the Universal Steam Oven, select "Single Point Stratified Grill" mode for 20 minutes, the dish is done.

中国大锅菜

The Big-Wok-Made Cuisine of China, Food Volume of Steam Oven（Commemorative Edition）

蒸烤箱卷（纪念版）

菜品特点

特色 这道菜是由传统宴会名菜『冬菜扒鸭』演变而来，是津门经典菜，整鸭切块，加上土豆与胡萝卜，便成为一道适合团膳的菜品。冬菜分津冬菜与川冬菜两种，津冬菜用白菜，川冬菜用芥菜，二者味道有所不同。本菜单选用色泽金黄、香气浓郁的津冬菜制作。

品味 津冬菜在腌制的过程中加入了蒜泥，具有十分浓郁的香味，非常适合用来烧制菜肴，味道鲜美，质地脆嫩，咸淡适口，回味甘甜。鸭块易熟，减少烹调时间可有效保持肉质的鲜嫩，口感极佳。

品相 料汁的颜色奠定了整道菜的品相，津冬菜颜色较浅，呈深黄色，加上酱油，菜品呈现酱黄色，从菜色可知鸭肉十分入味，视觉效果诱人。

营养价值 津冬菜以大白菜为主，含有丰富的维生素，具有增强抵抗力，促进消化，解渴利尿的作用。鸭肉中含有丰富的蛋白质，容易被人体吸收，所含B族维生素和维生素E较其他肉类多，能有效抵抗脚气病，神经炎和多种炎症，还能抗衰老。

菜品名称

酱爆笋丁

Name: Quick Deep-fried Winter Bamboo Shoots with Bean Paste

制作人：林　勇　　中国烹饪大师

Made by: Yong Lin　　A Great Master of Chinese Cuisine

主　料 Main Ingredient
冬　笋：1250g　切丁
Winter Bamboo Shoots　1250g
Chopped

配　料 Burdening
芹　菜：750g　切丁
Celery　750g　Chopped

胡萝卜：500g　切丁
Carrot　500g　Chopped

调　料 Seasoning
清　油 Oil 50ml
盐 Salt 13g
白　糖 Sugar.......................15g
甜面酱 Sweet Bean Paste 40g

料　酒 Cooking Wine 20ml
干黄酱 Dried Bean Paste......80g
胡椒粉 Ground Pepper 3g
香　油 Sesame Oil 60ml
淀　粉 Starch......................30g
姜　末 Minced Ginger 20g
料　油 Spicing Oil 20ml

中国大锅菜

The Big-Wok-Made Cuisine of China, Food Volume of Steam Oven（Commemorative Edition）

蒸烤箱卷（纪念版）

制作方法

❶ 将甜面酱 40g 和干黄酱 80g 倒入小盆中，加香油 30ml 拌匀，放入万能蒸烤箱选择『单点分层蒸煮』模式蒸 30 分钟，制成酱汁备用。

❷ 将冬笋丁和胡萝卜丁倒入盆中上底味，加盐 10g，姜末 10g，胡椒粉 3g，拌匀后，加香油 10ml，干淀粉 30g，不停搅拌，将 15ml 料油拌入。将芹菜倒入盆中上底味，加盐 3g，料油 5ml，拌匀后备用。

❸ 将上好底味的冬笋和胡萝卜放入万能蒸烤箱飞水，选择『单点分层蒸煮』模式，时长 5 分钟。

❹ 飞水时炒制料汁，锅热下油 30ml，加姜末 10g 爆香，倒入备用的酱汁，加料酒 20ml，香油 20ml，白糖 15g，煸炒透，料汁即成。

❺ 将芹菜丁倒入飞水后的冬笋和胡萝卜丁，均匀浇上料汁，放入万能蒸烤箱烹制入味，选择『单点分层煎烤』模式，时长 2 分钟，即可出锅。

WORKING PROCESS

1. Mix the sweet beans paste 40g with dried bean paste 80g into a basin, add sesame oil 30ml and stir well, and steam by using the Universal Steam Oven, select "Single Point Stratified Steam" mode for 30 minutes, set the paste sauce aside for later use.

2. Marinate the winter bamboo shoots pieces and carrot pieces in a basin. Add salt 10g, minced ginger 10g, ground pepper 3g, well stir and pour sesame oil 10ml, starch 30g, keep stirring, pour spicing oil 15ml. Marinate celery in a basin, add salt 3g, spicing oil 5ml, well stir for later use.

3. Quick steam the marinated winter bamboo shoots and carrot by using the Universal Steam Oven, select "Single Point Stratified Steam" mode for 5 minutes.

4. Make the sauce at the same time. Pour oil 30ml in a heated wok, add minced ginger 10g and deep-fry till fragrance, pour the sauce as well as cooking wine 20ml, sesame oil 20ml, sugar 15g, quick deep-fry and the sauce is done.

5. Mix the celery pieces with steamed winter bamboo shoots and carrot pieces, pour the sauce evenly, bake by using the Universal Steam Oven, select "Single Point Stratified Grill" mode for 2 minutes. The dish is done.

中国大锅菜

菜品名称 · 酱爆笋丁

Name: Quick Deep-fried Winter Bamboo Shoots with Bean Paste

菜品特点

特色 酱爆是老北京的传统烹调手法之一，衍生出很多菜肴，通过煸炒甜面酱和干黄酱调制成的酱汁，赋予食材以浓郁的酱香味，一直以来，深受广大食客欢迎。这道菜对炒制酱汁时的火候把握要求极高，火大则易糊，火小则香味不够。

品味 酱爆笋丁就是『酱爆』系的菜肴之一，菜品酱香浓郁，入口咸鲜，咸中有甜。用万能蒸烤箱制作此菜时，在飞水时可以有效锁住食材的营养成分，保证口感的脆爽。

品相 食材均要切丁，与酱汁结合后，仿佛每一粒食材都穿上了一层『酱衣』，也赋予了他们浓浓的酱香。

营养价值 冬笋是一种富有营养价值并具有医药功能的美味食品，质嫩味鲜，清脆爽口，含有蛋白质和多种氨基酸、维生素，以及钙、磷、铁等微量元素以及丰富的纤维素，能促进肠道蠕动，既有助于消化，又能预防便秘和结肠癌的发生。

菜品名称

南瓜粉蒸鸡

Name: Steamed Chicken with Pumpkin

制作人：林　勇　　中国烹饪大师

Made by: Yong Lin　　A Great Master of Chinese Cuisine

主　料 Main Ingredient　　　　Pumpkin　1250g　Chopped　　香　油 Sesame Oil 30ml
鸡腿肉：1000g　切块　　　　　　　　　　　　　　　　　　胡椒粉 Ground Pepper 6g
Drumstick　1000g　Pieced　　　**调　料 Seasoning**　　　米　粉 Rice Flour 100g
　　　　　　　　　　　　　　　甜面酱 Sweet Bean Paste ....15g　姜　末 Minced Ginger 20g
配　料 Burdening　　　　　盐 Salt ....20g　葱　末 Minced Scallion.........20g
南　瓜：1250g　切小块　　　　蚝　油 Oyster Oil.................10g

中国大锅菜

菜品名称 · 南瓜粉蒸鸡

Name: Steamed Chicken with Pumpkin

制 作 方 法

❶ 将鸡块放入盆中腌制入味，加盐 10g，胡椒粉 3g，蚝油 10g，甜面酱 15g，香油 15ml，拌匀后，加葱、姜末各 10g 拌匀，倒入米粉 100g，让米粉粘在鸡块上，腌制 15 分钟。

❷ 将南瓜放入盆中上底味，加盐 10g，胡椒粉 3g，葱、姜末各 10g，香油 15ml，拌匀。

❸ 将南瓜倒入烤盘铺平，将鸡块铺在南瓜上，放入万能蒸烤箱烹熟，选择『单点分层蒸煮』模式，时长 15 分钟。

WORKING PROCESS

1. Marinate the pieced drumstick in a basin, add salt 10g, ground pepper 3g, oyster sauce 10g, sweet bean paste 15g, sesame oil 15ml, well stir then add minced scallion 10g, minced ginger 10g, stir evenly, add rice flour 100g, coat it on the drumstick, marinate for 15 minutes.

2. Marinate pumpkin in a basin, add salt 10g, ground pepper 3g, minced scallion 10g, minced ginger 10g, sesame oil 15ml, stir evenly.

3. Place the pumpkin on the bottom of baking pan, place chicken on the pumpkin, steam by using the Universal Steam Oven, select "Single Point Stratified Steam" mode for 15 minutes.

中国大锅菜

蒸烤箱卷（纪念版）

The Big-Wok-Made Cuisine of China, Food Volume of Steam Oven (Commemorative Edition)

菜品特点

特色 南瓜原产于中美洲，明代传入我国。由于其颜色金黄，在福建、浙江沿海一带又称之为『金瓜』，目前在我国已经衍生出多个品种，南北方广泛种植，本道菜选用的南瓜品种为小磨盘，可谓型如其名。南瓜粉蒸鸡是一道南方地区的经典家常菜，江南地区还会加糯米一起蒸，则又为菜品增加了谷物的香甜。

品味 这道菜软嫩易嚼，十分适合老人和小孩食用，鸡肉细嫩清淡，南瓜绵软香甜，口感软糯，品味咸甜。

品相 鸡肉与南瓜均为黄色，视觉上就令人心生喜爱，给人以香甜可口的感觉，闻其香味，则更令人迫不及待，是一道色香味俱佳的菜品。

营养价值 南瓜营养丰富，除了有人体所需要的多种维生素外，还含有易被人体吸收的磷、铁、钙等多种营养成分，又有补中益气、消炎止痛、解毒杀虫的作用，对老年人高血压、冠心病、肥胖症等，亦有较好的疗效。鸡肉是蛋白质最高的肉类之一，是属于高蛋白低脂肪的食品，且易被人体吸收利用。此外，鸡肉还含有脂肪、钙、磷、铁、镁、钾、钠、维生素等营养成分。

菜品名称

油淋羊肉

Name: Pouring-Oil Mutton

制作人：林　勇　　中国烹饪大师

Made by: Yong Lin　　A Great Master of Chinese Cuisine

主　料　Main Ingredient
羊　肉：1500g　切条
Mutton　1500g　Shred

配　料　Burdening
白　菜：1500g　切片
Cabbage　1500g　Sliced

调　料　Seasoning
清　油　Oil...........................65ml

盐　Salt 30g
胡椒粉　Ground Pepper 8g
料　酒　Cooking Wine 10ml
淀　粉　Starch.....................45g
辣椒面　Chili Powder 20g
红椒碎　Chopped chili............10g
鸡　蛋　Egg...............1 个 (pcs)
葱　末　Minced Scallion.........10g
姜　末　Minced Ginger 10g
料　油　Spicing Oil 20ml

备　注　Tips
滑油时油温不要太高，因为羊肉不会
再进入万能蒸烤箱烹制，所以羊肉一
定要滑熟。
Make sure the mutton be cooked
during pouring oil, because the mutton
will not be cooked again by using the
Universal Steam Oven.

中国大锅菜

蒸烤箱卷（纪念版）

The Big-Wok-Made Cuisine of China, Food Volume of Steam Oven（Commemorative Edition）

制作方法

① 将羊肉放入盆中，加盐 15g，胡椒粉 5g，料酒 10ml。拌匀后打入 1 个鸡蛋，搅拌摔打，加干淀粉 15g 上浆，摔打上劲，倒入清油 15ml，腌制 15 分钟。将 30g 淀粉制成水淀粉备用。

② 将白菜放入盆中上底味，加盐 15g，胡椒粉 3g，葱、姜末各 10g，拌匀后浇上料油 20ml，倒入漏眼蒸盘中，放入万能蒸烤箱飞水，选择『单点分层蒸煮』模式，时长 3 分钟。

③ 将腌制好的羊肉放入油锅滑油。

④ 将滑油后的羊肉码在飞水后的白菜上，撒上辣椒面 20g，红椒碎 10g，锅内烧热油 50ml，将热油泼在羊肉上，菜品即成。

WORKING PROCESS

1. Put the mutton into basin, add salt 15g, ground pepper 5g, cooking wine 10ml, 1 egg, starch 15g to coat the mutton and knead. Pour oil 15ml and marinate for 15minutes. Turn the starch 30g into water-starch mash for later use.

2. Marinate the cabbage in a basin, add salt 15g, ground pepper 3g, minced scallion 10g, minced ginger 10g, well stir and pour spicing oil 20ml, quick steam by using the Universal Steam Oven, select "Single Point Stratified Steam" mode for 3 minutes.

3. Fry the mutton quickly.

4. Place the quick fried mutton on the quick steamed cabbage, sprinkle chili powder 20g, chopped red bell pepper 10g, heat oil 50ml, then pour the hot oil on the mutton, the dish is done.

中国大锅菜

菜品名称·油淋羊肉
Name: Pouring-Oil Mutton

菜品特点

特色 这是一道充满草原风情的菜肴，羊肉本为热性，和辣椒完美结合，火辣辣的感觉令人顿生暖意，是御寒佳肴。「油淋」是烹调技术中一种特殊的炸法，适用于质地鲜嫩的原料，它既发扬炸制香脆之长处，又保留原料鲜嫩之特色。

品味 羊肉本身带有膻味，辣椒可以有效遮盖住膻味，有助于打开食客的胃口。「油淋」之法则既保证了羊肉的鲜嫩，又将辣椒的干香爆出。羊肉的鲜香与辣椒的干香融合在一起，给人的味蕾以强烈刺激，令人食欲大增，而大白菜也起到了缓解辣味的作用。

品相 泼油使菜品发出「呲呲」之声，辣椒的颜色也瞬间鲜艳起来，红嫩的羊肉与火红的辣椒有如草原人民的热情，令人心中顿生喜悦。

营养价值 羊肉与辣椒均为热性食品，这道菜堪称冬季滋补佳品，是寒冷地区人们抵御严寒的最爱。羊肉含有丰富的蛋白质、脂肪，同时还含有维生素B₁、B₂及矿物质钙、磷、铁、钾、碘等，营养全面、丰富。白菜含有多种营养物质，是人体生理活动所必需的维生素、无机盐及食用纤维素的重要来源，并含有丰富的钙，是预防癌症、糖尿病和肥胖症的健康食品。

肉丝炒豆芽

菜品名称

香 酥 鸭

———— ❧ ————

Name: Crispy Fried Duck

制作人：苏喜斌　　中国烹饪大师

Made by: Xibin Su　　A Great Master of Chinese Cuisine

主　料 Main Ingredient	桂　皮 Cassia......................30g	生　抽 Light Soy Sauce......30ml
樱桃谷鸭：4000g	八　角 Anise 15g	淀　粉 Starch.......................50g
Cherry Valley Ducks　4000g	香　油 Sesame Oil 25ml	鸡　蛋 Egg.................3 个 (pcs)
	花　椒 Chinese Prickly Ash 5g	葱　段 Scallion 20g
调　料 Seasoning	料　酒 Cooking Wine 50ml	姜　片 Sliced Ginger 20g
盐　Salt 40g	胡椒粉 Ground Pepper 5g	

中国大锅菜

菜品名称·香酥鸭

Name: Crispy Fried Duck

制作方法

❶ 将鸭子去除内脏和喉管，对半切成两片，将鸭子的骨骼展开，揉制疏松，使其更容易腌制入味。

❷ 腌制鸭肉，加桂皮30g，八角15g，香油5ml，葱段20g，姜片20g，花椒5g，料酒50ml。将30g盐均匀抹在表面，加胡椒粉5g，倒入生抽30ml，拌制均匀。将调料在鸭片上搓一搓，让料汁均匀覆盖在鸭身上，并且起到了按摩的作用，然后腌制90分钟。

❸ 将腌制好的鸭肉放入万能蒸烤箱蒸熟，选择『蒸制蔬菜』模式，时长30分钟。

❹ 蒸鸭子时调配酥糊，碗内打入3个鸡蛋清打散，加盐10g，淀粉50g，淀粉和蛋清要拌匀，不要有颗粒，再加入20ml香油。

❺ 将酥糊均匀抹在蒸好的鸭子表面，淋上少许明油，放入万能蒸烤箱炸制，选择『单点分层炙烤』模式，时长8分钟，菜品即成。

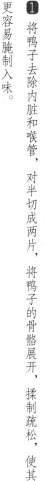

WORKING PROCESS

1. Remove the duck's offal and throat tube, cut into two pieces half-and-half, spread the duck bone, knead and soften it, make it to be easier to pickle flavor.

2. Marinate the duck. Add cassia 30g, anise 15ml, sesame oil 5g, scallion 20g, sliced ginger 20g, Chinese prickly ash 5g, cooking wine 50ml, paint salt 30g on the duck, add ground pepper 5g, light soy sauce 30ml, stir evenly, and marinate for 90 minutes.

3. Steam the marinated duck by using the Universal Steam Oven, select "Steam Vegetable" mode for 30 minutes.

4. Make the starchy mash at the same time, break 3 eggs white in the bowl, add salt 10g, starch 50g, well stir and pour sesame oil 20ml.

5. Paint the starchy mash on the duck, pour a little oil, fry by using the Universal Steam Oven, select "Single Point Stratified Grill" mode for 8 minutes. The dish is done.

中国大锅菜

蒸烤箱卷（纪念版）

The Big-Wok-Made Cuisine of China, Food Volume of Steam Oven（Commemorative Edition）

菜品特点

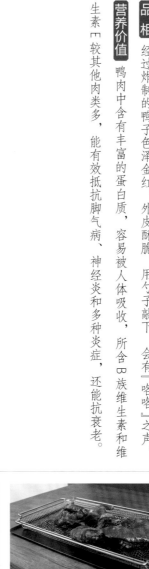

特色 这是四川名菜之一，皮酥肉嫩，曾经是招待外宾时的经典菜肴，亦广受好评。这道菜食材易得，制作简单，亦无川菜麻辣特点，更为人们所接受。用传统烹饪方法制作此菜最后要用油炸，火候较难掌握，若油温低炸不酥，肉也会渗油进去，很不好吃，但油温高了容易焦，味道发苦。用万能蒸烤箱制作此菜，可以更为精准地控制火候，做到恰到好处。

品味 香酥鸭的特点就在『香酥』两字上。香——香味浓郁，香气扑鼻；酥——酥软爽口，酥而不油。由于鸭子体积较大，要做到入味，一定要上足调味料，腌制较长时间，肉厚的地方要反复揉搓，使肉质松软，更易入味。

品相 经过炸制的鸭子色泽金红，外皮酥脆，用勺子敲下，会有『咯咯』之声。

营养价值 鸭肉中含有丰富的蛋白质，容易被人体吸收，所含B族维生素和维生素E较其他肉类多，能有效抵抗脚气病、神经炎和多种炎症，还能抗衰老。

椒 盐 虾

❦

Name: Spicy Salt Prawns

制作人：苏喜斌　　中国烹饪大师

Made by: Xibin Su　　A Great Master of Chinese Cuisine

主　料　Main Ingredient
白　虾：3000g　去虾线
Prawn　3000g　Cleaned

配　料　Burdening
青　椒：150g　切丝
Green Bell Pepper
150g　Shredded

红　椒：150g　切丝
Red Bell Pepper　150g　Shredded
洋　葱：100g
Onion　100g

调　料　Seasoning
清　油　Oil 120ml
料　油　Spicing Oil 20ml

胡椒粉 Ground Pepper 10g
淀　粉　Starch 30g
盐　Salt 30g
料　酒　Cooking Wine 70ml
料　油　Spicing Oil 20ml

中国大锅菜

The Big-Wok-Made Cuisine of China, Food Volume of Steam Oven（Commemorative Edition）

蒸烤箱卷（纪念版）

制作方法

❶ 将去好虾线的白虾入盆腌制，加入盐 10g，胡椒粉 5g，淀粉 30g，料酒 20ml，搅拌均匀，淋上清油 20ml 拌匀，腌制 10 分钟。

❷ 将腌制好的白虾倒入炸筐，放入万能蒸烤箱炸制，选择『单点分层炙烤』模式，时长 5 分钟。

❸ 炸虾时炒制椒盐料汁，锅热下油 100ml，加洋葱炒香，加入青、红椒丝，倒入料酒 50ml，加盐 20g，胡椒粉 5g，浇上料油 20ml，料汁即成。

❹ 将料汁倒入炸好的虾中，搅拌均匀，菜品即成。

WORKING PROCESS

1. Clean the prawn and marinate it. Add salt 10g, ground pepper 5g, starch 30g, cooking wine 20ml, well stir, pour oil 20ml, marinate for 10 minutes.

2. Bake the marinated prawn by using the Universal Steam Oven, select "Single Point Stratified Grill" mode for 5 minutes.

3. Make the spicy salt sauce at the same time. Pour oil 100ml into a heated wok, add onion and stir-fry till fragrance, add stripped green and red bell pepper, cooking wine 50ml, salt 20g, ground pepper 5g, pour spicing oil 20ml, the sauce is done.

4. Pour the sauce in fried prawn, well stir and the dish is done.

中国大锅菜

菜品名称·椒盐虾
Name: Spicy Salt Prawns

菜品特点

特色 这是一道曾经在中国台湾盛极一时的菜肴，它源于台南，随着台湾小吃的流行，胡椒虾也广受人们的喜爱。今天，这道菜肴已经十分普遍。用万能蒸烤箱制作此菜，可以将步骤分解为腌制、炸熟、浇汁入味三步，既方便快捷，又能保证香浓的胡椒味。

品味 虾肉肉质鲜美，在虾皮的表面加较重口味的调味料，品尝此菜，初觉胡椒的特殊香味，吃到虾肉，又为它的白嫩鲜美所陶醉。胡椒既能去除腥味，又起到调味的作用，夸张的口味对比满足了人们的需求，亦将鲜美衬托出来。喜欢吃虾头的朋友千万不要错过，虾黄的鲜美更胜过白灼。

品相 这道菜色泽红亮，挂糊炸后的虾表面挂有细小淀粉颗粒，充分吸收了料汁，颜色诱人。

营养价值 虾类营养丰富，且肉质松软，易消化，对身体虚弱以及病后需要调养的人是极好的食物；虾中含有丰富的镁，镁对于心脏活动具有重要的调节作用，且能够很好地保护心血管系统。

菜品名称

酱爆鸡丁

Name: Stir-fried Chicken Breast with Soy Bean Paste

制作人：苏喜斌　　中国烹饪大师

Made by: Xibin Su　　A Great Master of Chinese Cuisine

主 料 Main Ingredient
鸡胸肉：2500g 切丁
Chicken Breast　2500g　Pieced

配 料 Burdening
胡萝卜：750g 切丁
Carrot　750g　Pieced
黄 瓜：750g 切丁

Cucumber　750g　Pieced

调 料 Seasoning
清 油 Oil..........................120ml
盐 Salt 20g
料 油 Spicing Oil 20ml
甜面酱 Sweet Bean Paste ..200g
酱 油 Soy Sauce 45ml

老 抽 Dark Soy Sauce........6ml
淀 粉 Starch.......................50g
料 酒 Cooking Wine 30ml
白 糖 Sugar........................15g
葱 末 Minced Scallion........10g
姜 末 Minced Ginger 10g

中国大锅菜

菜品名称 · 酱爆鸡丁

Name: Stir-fried Chicken Breast with Soy Bean Paste

制作方法

❶ 首先将鸡丁腌制上浆，将鸡丁放入盆中，加盐10g，料酒10ml，甜面酱80g，老抽3ml，酱油15ml，搅拌均匀，淋上清油20ml，倒入淀粉30g上浆，打入适量水分，使鸡丁更加鲜嫩。

❷ 将鸡丁放入万能蒸烤箱滑油，选择『单点分层炙烤』模式，时长3分钟。将黄瓜丁和胡萝卜丁放入万能蒸烤箱飞水，选择『蒸制蔬菜』模式，黄瓜时长1分钟，胡萝卜4分钟。

❸ 烧制料料汁，锅热下油100ml，加葱、姜末各10g爆香，加入甜面酱120g炒出香味，加白糖15g，将白糖炒化，顺锅边加入料酒20ml，使料酒香味散发出来，再加入酱油30ml，老抽3ml，盐10g，倒入20g淀粉制成的水淀粉勾芡，锅开后淋上料油20ml，料汁即成。

❹ 将飞水后的蔬菜倒入鸡丁中，浇上料汁拌匀，放入万能蒸烤箱炒制入味，选择『单点分层煎烤』模式，时长3分钟。

WORKING PROCESS

1. Marinate the chicken. Put the chicken in a basin, add salt 10g, cooking wine 10ml, sweet bean paste 80g, dark soy sauce 3ml, soy sauce 15ml, well stir, pour oil 20ml, pour starchy mash to coating, add moderate water.

2. Quick fry the chicken by using the Universal Steam Oven, select "Single Point Stratified Grill" mode for 3 minutes. Quick steam the chopped cucumber and carrot by using the Universal Steam Oven, select "Steam Vegetable" mode, 1 minute for cucumber and 4 minutes for carrot.

3. Make the sauce. Pour oil 100ml into a heated wok, add minced scallion 10g, minced ginger 10g, deep-fry till fragrance, then add sweet bean paste 120g, sugar 15g, cooking wine 20ml, soy sauce 30ml, dark soy sauce 3ml, salt 10g, pour water-starch mash 20g to thicken the sauce, then pour spicing oil 20ml, the sauce is done.

4. Mix the quick steamed vegetable with chicken, pour sauce and stir well, sir-fry by using the Universal Steam Oven, select "Single Point Stratified Bake" mode for 3 minutes.

中国大锅菜

蒸烤箱卷（纪念版）

The Big-Wok-Made Cuisine of China, Food Volume of Steam Oven（Commemorative Edition）

菜品特点

特色 老北京的甜面酱是十分有名的，也衍生出来了很多使用甜面酱烹调的菜肴，酱爆系列的菜品就是其中之一。早些年北京的餐馆流行酱爆肉丁，以猪肉为主料，但是肉质较老，后来有些厨师受宫保鸡丁的启发，用鸡肉烹调此菜，成就了一道酱爆经典菜——『酱爆鸡丁』。

品味 鸡肉在腌制时可以打入适量水分，这样滑油后更加鲜嫩，口感奇佳。酱爆法烹制的菜肴酱香浓郁，入口咸鲜，咸中有甜，可夹饼而食，无比美味。

品相 这道菜色泽红亮，鸡丁和蔬菜都挂满浓稠的芡汁，酱色较深，传递着菜品的品位。

营养价值 鸡肉性平、温、味甘，入脾、胃经，可益气、补精、添髓。而鸡翅中的胶原蛋白含量更加丰富，对于保持皮肤光泽、增强皮肤弹性均有好处。

大葱烧鱼段

Name: Braised Long Li Fish Fillets with Scallion

制作人：苏喜斌　　中国烹饪大师

Made by: Xibin Su　　A Great Master of Chinese Cuisine

主　料 Main Ingredient
龙俐鱼：4000g　切段
Long Li Fish　4000g　Cut

调　料 Seasoning
清　油 Oil.........................150ml
盐 Salt 40g
醋 Vinegar..........................80ml

胡椒粉 Ground Pepper.........18g
酱　油 Soy Sauce 50ml
白　糖 Sugar.......................20g
淀　粉 Starch......................80g
料　油 Spicing Oil 50ml
蚝　油 Oyster Oil...............50ml
姜　片 Sliced Ginger............30g
葱　段 Scallion 200g

备　注 Tips
倒入水淀粉时要慢慢倒入，边倒边搅拌，让料汁更加浓稠、均匀，挂在鱼上，味道更佳。
Pouring water-starch mash should be slow, with stirring, which will help to coat the fish evenly.

制作方法

❶ 首先腌制，将鱼放入盆中，加盐20g，胡椒粉8g，淀粉50g，水少许，用揉拌的方法将调料拌匀，腌制15分钟。

❷ 烤盘内刷油，将腌制好的龙俐鱼摆入盘中进行飞水处理，选择『单点分层煎烤』模式，时长3分钟。取30g淀粉加水制成水淀粉备用。

❸ 然后炒制料汁，锅热下油150ml，油热加姜片30g，葱段200g，大火炒香，将葱焗软，再加醋80ml，蚝油50g，酱油50ml，白糖20g，盐20g，胡椒粉10g。锅开后倒入水淀粉，待锅再次烧开，淋上料油50ml，制成料汁。

❹ 将料汁浇在飞水后的鱼肉上，再次放入万能蒸烤箱进行烹制，选择『单点分层煎烤』模式，时长3分钟，出锅即可盛盘。

WORKING PROCESS

1. Marinate the fish at first. Put the fish fillets in a basin. Add salt 20g, ground pepper 8g, starch 50g, a little water, knead and stir evenly. Marinate for 15 minutes.

2. Brush oil at the baking pan, place the marinated Long Li fish fillets in the pan and quickly frying, select "Single Point Stratified Grill" mode for 3 minutes. Turn starch 30g into water-starch mash for later use.

3. Make the sauce. Pour oil 150ml into a heated wok, add sliced ginger 30g, cut scallion 200g, deep-frying till fragrance, pour vinegar 80ml, oyster oil 50g, soy sauce 50ml, sugar 20g, salt 20g, ground pepper 10g, and add water-starch mash while the soup is boiled, then pour spicing oil 50ml.

4. Pour the sauce on the quick fried fish fillets, bake by using the Universal Steam Oven again, select "Single Point Stratified Grill" mode for 3 minutes. Then the dish is done.

中国大锅菜

菜品名称·大葱烧鱼段

Name: Braised Long Li Fish Fillets with Scallion

菜品特点

特色 葱烧是鲁菜的著名烹调手法，以传统名菜『葱烧海参』为代表，这道菜与其相似。龙俐鱼是物美价廉的海产鱼类，鱼肉呈块状，有着海鱼的鲜美而腥味较淡，且只有中间一根主刺，十分适合团餐使用。

品味 龙俐鱼松软质嫩，鲜香无比，而大葱起到了重要的调味作用，不但能去除水产品中的腥味，还能产生特殊的香味，配合上酱油、蚝油，更加突出鱼肉的鲜美。

品相 一节节雪白的大葱是这道菜的名片，辅以各种调味料，菜色呈现出深红的酱色，鱼肉挂满料汁，香味诱人。

营养价值 龙俐鱼具有海产鱼类在营养上显著的优点，含有较高的不饱和脂肪酸，蛋白质容易消化吸收。其肌肉肉细嫩，口感爽滑，鱼肉久煮不老，无腥味和异味，属于高蛋白、低脂肪、富含维生素的鱼类。

粉皮炖鱼

Name: Stewed the Long Li Fish with Sheet Jelly

制作人：孙家涛　　中国烹饪大师

Made by: Jiatao Sun　　A Great Master of Chinese Cuisine

主　料 Main Ingredient
龙俐鱼：2000g　切块
Long Li Fish　2000g　Cut

配　料 Burdening
粉　皮：1000g　泡发切段
Sheet Jelly　1000g　Chopped

调　料 Seasoning
清　油 Oil 50ml
盐　Salt 30g
酱　油 Soy Sauce 30ml
醋　Vinegar........................30ml
料　酒 Cooking Wine 30ml
白　糖 Sugar......................50g

干辣椒 Dried Chili 5g
八　角 Anise 5g
淀　粉 Starch......................60g
姜　末 Minced Ginger 20g
蒜　片 Sliced Garlic.............10g
葱　末 Chopped Scallion 20g

中国大锅菜

菜品名称·粉皮炖鱼

Name: Stewed the Long Li Fish with Sheet Jelly

制作方法

① 将粉皮放在水中泡发。将30g淀粉加水制成水淀粉备用。

② 炒糖色，锅热倒入油20ml，倒入糖50g，炒糖色，成枣红色，浇上水150ml，糖色即成。

③ 炒制料汁，锅热下油30ml，油热加葱花20g，姜末20g，蒜片10g，八角5g，炒香，加干辣椒段5g，料酒30ml，醋30ml，酱油30ml，倒入水750ml，水开后加盐20g，倒入糖色，加水淀粉勾芡，芡汁要薄，料汁即成。

④ 将龙俐鱼加10g盐，拍生粉30g，放入万能蒸烤箱滑油，烤盘内刷底油，将拍粉后的鱼摆入盘中，刷一层明油，选择「单点分层炙烤」模式，时长3分钟。

⑤ 将粉皮放入鱼中，浇上料汁，搅拌均匀，放入万能蒸烤箱继续烹制，选择「单点分层煎烤」模式，时长3分钟，即可出锅。

WORKING PROCESS

1. Soaking the sheet jelly in water. Turn the starch 30g into water-starch mash for later use.

2. Make the caramel. Pour oil 20ml into a heated wok, add sugar 50g, stir-fry till turns into heavy red, pour water 150ml, the caramel is done.

3. Make the sauce. Pour oil 30ml into a heated wok, then add chopped scallion 20g, minced ginger 20g, sliced garlic 10g, anise 5g, stir-fry till fragrance, add dried chili 5g, cooking wine 30ml, vinegar 30ml, soy sauce 30ml, pour water 750ml, add salt 20g while the water boiled, pour caramel, thicken the sauce with water-starch mash, but don't too heavy, the sauce is done.

4. Salt Long Li fish, coat with starch 30g. Stir-fried by using the Universal Steam Oven, brush oil at the bottom of the baking pan, select "Single Point Stratified Grill" mode for 3 minutes.

5. Put the sheet jelly into the fish, dress with sauce and well stirred, baked by using the Universal Steam Oven, select "Single Point Stratified Grill" mode for 3 minutes. The dish is done.

中国大锅菜

蒸烤箱卷（纪念版）

The Big-Wok-Made Cuisine of China, Food Volume of Steam Oven（Commemorative Edition）

菜品特点

特色 这是一道地道的东北风味菜，此菜对鱼的种类要求不高，新鲜即可。东北严寒，冬季河湖冻住，渔民凿冰捕鱼，这是粉皮炖鱼的上品，沿海地区可以用海鱼，味道亦十分鲜美。本菜用的是龙俐鱼，肉质细嫩，鱼刺很少，方便作为团餐菜品食用。粉皮起源于中国，最早记载于北魏年间的农书齐民要术中，与很多食材成为中华美食的经典搭配。

品味 粉皮炖鱼以酱香味为主，由于有糖色，既提升了鱼肉的鲜味，又使口味上多了一点淡甜之味，咸鲜适口。龙俐鱼肉质嫩少刺，十分便于食用，粉皮经过炖制，将料汁充分吸收，充分保留了鱼肉的鲜香和料汁的酱香，口感软糯，入口香甜。

品相 龙俐鱼本身洁白如雪，滑油后通体金黄，加料汁炖制，挂上一层薄薄的酱汁，料汁中有酱油和糖色，整个菜品呈酱香之色，又不掩鱼肉的鲜美。粉皮晶莹剔透，充分吸收了料汁，滑爽而入味，小心翼翼夹起，一滑就到了嘴里。

营养价值 龙俐鱼具有海产鱼类在营养上显著的优点，含有较高的不饱和脂肪酸，蛋白质容易消化吸收。其肌肉细嫩，口感爽滑，鱼肉久煮不老，无腥味和异味，属于高蛋白、低脂肪、富含维生素的鱼类。粉皮的重要营养成分是碳水化合物，还含有少量蛋白质、维生素及矿物质，十分易于被人体吸收，补充身体所需能量。

菜品名称

虾仁熘水蛋

Name: Steamed Eggs with Shrimps

制作人：孙家涛　　中国烹饪大师

Made by: Jiatao Sun　　A Great Master of Chinese Cuisine

主 料 Main Ingredient
虾 仁：1000g
Shrimp　1000g

鸡 蛋：1000g
Egg　1000g

配 料 Burdening
青、红椒：100g 切碎
Green/Red Bell Pepper　100g
Chopped

调 料 Seasoning
清 油 Oil........................250ml

淀 粉 Starch......................60g
白 糖 Sugar........................10g
盐 Salt 40g
料 酒 Cooking Wine 15ml
姜 末 Minced Ginger 20g
蒜 片 Sliced Garlic..............15g
葱 花 Chopped Scallion 30g

中国大锅菜

蒸烤箱卷（纪念版）

The Big-Wok-Made Cuisine of China, Food Volume of Steam Oven (Commemorative Edition)

制作方法

❶ 虾仁上浆：虾仁放入盆中，打入两个鸡蛋的蛋清，加盐15g，料酒15ml，淀粉30g，用搅拌的方法将调料拌匀，放入冰箱腌制15分钟。

❷ 将鸡蛋加盐10g打散，倒入清油200ml备用。将30g淀粉加水制成水淀粉备用。

❸ 烧制料汁：锅热下油50ml，加葱花30g，姜末20g，蒜片15g炒香。加青、红椒碎100g，倒入水700ml，加盐15g，糖10g。烧开后倒入水淀粉勾芡，加香油10ml，料汁既成。

❹ 将腌制好的虾仁进行飞水处理，选择『单点分层煎烤』模式，时长2分钟。

❺ 将烤盘刷底油，撒入飞水后的虾仁，倒上鸡蛋液拌匀，选择『单点分层煎烤』模式，时长3分钟。

❻ 将出锅后的鸡蛋虾仁浇上料汁，搅拌均匀，即可盛盘。

WORKING PROCESS

1. Coat shrimps with starch. Put the shrimps into a basin, add 2 eggs' egg-white, salt 15g, cooking wine 15ml, starch 30g, stir the seasoning evenly, marinate for 15 minutes in the fridge.

2. Stir the eggs with salt 10g, pour oil 200ml for later use. Make starch 30g to be water-starch mash for later use.

3. Make the sauce. Pour oil 50ml into a heated wok, add chopped scallion 30g, minced ginger 20g, sliced garlic 15g, stir-fried till fragrance, add chopped green/red bell pepper 100g, water 700ml, salt 15g, sugar 10g, thicken the sauce with water-starch mash after the water boiled, then add sesame oil 10ml, the sauce is done.

4. Bake the marinated shrimps by using the Universal Steam Oven, select "Single Point Stratified Bake" mode for 2 minutes.

5. Brush the oil at the bottom of the baking pan, sprinkle the steamed shrimps, pour the egg juice and stirred evenly, select "Single Point Stratified Bake" mode for 3 minutes.

6. Dress sauce, well stir, then the dish is done.

中国大锅菜

菜品名称·虾仁熘水蛋

Name: Steamed Eggs with Shrimps

菜品特点

特色 虾仁熘水蛋是一道简单、易做的家常菜，味道鲜美，由于口感滑嫩，十分受老人和小孩喜爱。

品味 虾仁味道十分鲜美，与水蛋一起蒸制，其鲜味与鸡蛋的香味融合，成为至鲜至香的一道菜品，且二者皆为质嫩之物，便于咀嚼，口感上佳。

品相 水蛋呈金黄之色，上面铺满一层红白相间的虾仁，菜色具有富贵之气，表面一层薄薄的芡汁，仿佛将鲜味都锁在里面。

营养价值 虾肉营养价值丰富，性温，富含蛋白质，而脂肪含量较低，易于消化，是滋补的佳品。鸡蛋几乎含有人体必需的所有营养物质，如蛋白质、脂肪、卵黄素、卵磷脂、维生素和铁、钙、钾，被人们称作『理想的营养库』。

油爆里脊丁

制作人：孙家涛　　中国烹饪大师

Made by: Jiatao Sun　　A Great Master of Chinese Cuisine

Name: Deep-fried Diced Pork Tenderloin

主 料 Main Ingredient
猪通脊：1500g 切丁
Pork Tenderloin　1500g　Diced

配 料 Burdening
黄 瓜：500g 切丁
Cucumber　500g　Diced

胡萝卜：500g 切丁
Carrot　500g　Diced

调 料 Seasoning
清 油 Oil......................100ml
盐 Salt 30g
白 醋 White Vinegar..........10ml

胡椒粉 Ground Pepper 5g
料 酒 Cooking Wine 90ml
白 糖 Sugar.......................20g
香 油 Sesame Oil 20ml
淀 粉 Starch....................110g
姜 末 Minced Ginger..........15g
葱 末 Chopped Scallion 15g

中国大锅菜

菜品名称·油爆里脊丁
Name: Deep-fried Diced Pork Tenderloin

制作方法

① 首先将里脊丁上浆腌制，里脊丁入盆。打入3个鸡蛋清，拌匀，加料酒60ml，盐10g，淀粉80g，摔打上劲，再加油50ml，继续摔打搅拌，腌制15分钟。将30g淀粉加水制成水淀粉备用。

② 将腌制好的里脊丁滑油，里脊丁要变色、熟透，捞出控油。将胡萝卜和黄瓜进行飞水，选择『蒸制蔬菜』模式，胡萝卜时长3分钟，黄瓜时长1分钟。

③ 炒制料汁，锅热下底油50ml，加葱末、姜末各15g，料酒30ml，白醋10ml，倒入水200ml，加糖20g，盐20g，胡椒粉5g，锅开后倒入水淀粉勾芡，淋上香油20ml，料汁即成。

④ 将里脊丁和飞水后的胡萝卜黄瓜倒在一起，均匀浇上料汁，搅拌均匀，放入万能蒸烤箱再次烹制，选择『单点分层煎烤』模式，时长3分钟，即可出锅。

WORKING PROCESS

1. Marinate the pork tender loin. Dice the pork tenderloin and put it into a basin. Break 3 eggs white, stir evenly, add cooking wine 60ml, salt 10g, flour 80g, knead and stir, pour oil 50ml, then continue to knead and marinate for 15 minutes. Make the starch 30g to be water-starch mash for later use.

2. Stir-fry the marinated pork tenderloin till the pork has been deeply fried. Take it out of the wok. Steam the cucumber and carrot by using the Universal Steam Oven, select "Steam Vegetable" mode, 3 minutes for carrot and 1 minute for cucumber.

3. Make the sauce, pour oil 50ml into a heated wok, add chopped scallion 15g, minced ginger 15g, cooking wine 30ml, white vinegar 10ml, pour water 200ml, add sugar 20g, salt 20g, ground pepper 5g, then thicken the sauce with water-starch mash after the water boiled, sprinkle sesame oil 20ml, oil 20ml, the sauce is done.

4. Mix the pork tenderloin and steam cucumber/carrot together, pour the sauce, stir it and balance seasonings, then bake by using the Universal Steam Oven, select "Single Point Stratified Bake" mode for 3 minutes, the dish is done.

中国大锅菜

The Big-Wok-Made Cuisine of China, Food Volume of Steam Oven (Commemorative Edition)

蒸烤箱卷（纪念版）

菜品特点

特色 这是一道北京菜，里脊肉质较嫩，但在烹制过程中容易变老，肉质会较硬，需要事先进行腌制处理，滑油时对火候的要求也比较高。制作方法看似简单，然而腌制用料、手法和滑油火候的把握却十分讲究，如此方能制作一道上好的油爆里脊丁。

品味 里脊是猪肉中最为软嫩的瘦肉部分，又有着浓郁的肉香，是这道菜最为核心的部分，各种调料在其中起到帮衬的作用，将里脊的味道凸显出来，又遮掩掉肉类中的异味。胡萝卜和黄瓜口感脆爽，作为里脊的调剂配菜，恰到好处。

品相 里脊丁经过滑油会变色，略有发白，与胡萝卜和黄瓜搭配，看上去略显寡淡，但是淡淡的颜色中却深藏浓郁的肉香味，让人食之欲罢不能。

营养价值 猪脊肉含有人体所需的丰富的优质蛋白、脂肪、维生素等，而且肉质较嫩，易消化，能为人类提供优质蛋白质和必需的脂肪酸，中医认为其具有补肾养血，滋阴润燥的作用。

芫爆里脊丝

Name: Deep-fried Stripped Pork Tenderloin

制作人：孙家涛　　中国烹饪大师

Made by: Jiatao Sun　　A Great Master of Chinese Cuisine

主　料　Main Ingredient

猪里脊：2000g　切丝
Pork Tenderloin　2000g　Stripped

配　料　Burdening

香　菜：150g　切段
Caraway　150g　Cut

调　料　Seasoning

清　油	Oil	250ml
盐	Salt	50g
白　醋	White Vinegar	50ml
胡椒粉	Ground Pepper	10g
料　酒	Cooking Wine	90ml
鸡　蛋	Egg	5 个 (pcs)
淀　粉	Starch	80g
姜　丝	Stripped Ginger	20g
葱　丝	Stripped Scallion	30g

备　注　Tips

腌制好的里脊丝放入冰箱冰镇 4 小时，效果会更好。

The marinated pork should be frozen in fridge for 4 hours before cooking, which will make it more delicious.

中国大锅菜

蒸烤箱卷（纪念版）

The Big-Wok-Made Cuisine of China, Food Volume of Steam Oven（Commemorative Edition）

制作方法

❶ 首先将里脊丝上浆腌制，里脊丝入盆，打入5个鸡蛋蛋清，加料酒30ml，盐20g，淀粉80g，油100ml，搅拌均匀，腌制15分钟。

❷ 调制料汁：盆内倒入白醋50ml，料酒60ml，盐30g，胡椒粉10g，倒入油100ml，搅拌均匀，加入料汁即成。

❸ 将腌制好的里脊丝滑油，里脊丝要变色、熟透。

❹ 准备2个烤盘，倒入底油50ml，放入万能蒸烤箱加热，选择『单点分层炙烤』模式，时长3分钟。

❺ 油热出盘，每盘撒上葱丝30g，姜丝20g，将滑油后的里脊丝均匀倒入热油中，放入万能蒸烤箱烹制，选择『单点分层炙烤』模式，时长1分钟。

❻ 出锅后，撒上香菜段，即可盛盘。

WORKING PROCESS

1. Marinate the pork tenderloin strips. Put them into a basin. Break 5 eggs' egg-white, pour cooking wine 30ml, salt 20g, starch 80g, oil 100ml, stir evenly and marinate for 15 minutes.

2. Make the sauce, pour white vinegar 50ml, cooking wine 60ml, salt 30g, ground pepper 10g, oil 100ml, stir evenly, the sauce is done.

3. Stir-fry the pork tenderloin strips till its color turns.

4. Prepare 2 baking pans, pour oil 50ml, grill by using the Universal Steam Oven, select "Single Point Stratified Grill" mode for 3 minutes.

5. Sprinkle stripped scallion 30g in each baking pan after well grilled, and add stripped scallion 30g, stripped ginger 20g. Pour the stir-fried stripped pork tenderloin into boiled oil, continue to bake by using the Universal Steam Oven, select "Single Point Stratified Grill" mode for 1 minute.

6. Then sprinkle caraway, the dish is done.

中国大锅菜

菜品名称·芫爆里脊丝

Name: Deep-fried Stripped Pork Tenderloin

椒盐白虾

烧二冬

菜品特点

特色 这是一道鲁菜名菜，制作方法与油爆类似，滑油之后大火爆炒，出锅前加入香菜。

「芫」是「芫荽」，即是香菜，其中含有许多挥发油，其特殊的香气，就是挥发油散发出来的。它能祛除肉类的腥膻味，因此在一些菜肴中加些香菜，便能起到祛腥膻、增味道的独特功效。

品味 里脊丝经过腌制和用香菜烹调，其中的异味已经毫无踪影，只留下令人十分沉醉的肉香味，口味咸鲜，入口软嫩，香味让人着迷，久久回荡在口中。

品相 这道菜菜色格调雅致，色彩翠绿鲜艳，熟透的肉丝和略有生鲜气息的香菜，一生一熟，视觉上给这道菜以强烈冲撞，口味上也起到了调剂作用。

营养价值 猪肉含有丰富的优质蛋白质和必需的脂肪酸，并提供血红素和促进铁吸收的半胱氨酸，能改善缺铁性贫血；具有补肾养血，滋阴润燥的功效。猪里脊肉含有丰富的优质蛋白、脂肪、胆固醇含量相对较少，一般人群都可食用。香菜提取液具有显著的发汗清热透疹的功能，其特殊香味能刺激汗腺分泌，促使机体发汗，透疹。此外还具有和胃调中的功效，皆是因香菜辛香升散，能促进胃肠蠕动，具有开胃醒脾的作用。

菜品名称

冬菜扒大鸭

Name: Grilled Duck with Sichuan Preserved Leaf Mustard

制作人：孙立新　　中国烹饪大师

Made by: Lixin Sun　　A Great Master of Chinese Cuisine

主　料　Main Ingredient	调　料　Seasoning		备　注　Tips
整　鸭：1750g	酱　油　Soy Sauce 70ml		1. 宴会适合用整鸭，团餐可以将鸭腿肉切成块制作。
Duck　1750g	清　油　Oil........................ 130ml		2. 炒汁时料酒可以多放些，让鸭肉味道更加浓郁。由于烹制时间较长，酒精挥发，不会出现酒精味。
	香　油　Sesame Oil 100ml		
配　料　Burdening	泡　椒　Soaked chili 20g		1. The uncut duck will be preferred in banquet and the sliced duck meat will be preferred in staff dining hall.
猪里脊：300g　切丝	料　酒　Cooking Wine 150ml		
Pork Belly　300g　Shred	白　糖　Sugar....................... 20g		2. The portion of cooking wine can be added a little more to make the duck more tasty.
冬　菜：250g　切末	胡椒粉　Ground Pepper 5g		
Sichuan Preserved Leaf Mustard	盐　Salt 10g		
250g　Minced	淀　粉　Starch....................... 20g		
	姜　片　Sliced Ginger 50g		
	葱　段　Scallion 150g		

制作方法

❶ 首先将鸭子入盆，加酱油 50ml 上色，放入烤盘进入万能蒸烤箱烤制。选择『单点分层炙烤』模式，时长 5 分钟。

❷ 烤制鸭子时炒制料汁，锅热下油 130ml，油温在六成热时加里脊丝炒熟，加葱段 150g、姜片 50g，煸炒出香味。加入冬菜，浇上香油 80ml，泡椒 20g，料酒 150ml，酱油 20ml，白糖 20g，胡椒粉 5g，倒入水 3L，锅开后加盐 10g，浇上料汁即成。

❸ 将料汁倒入烤制好的鸭子，放入万能蒸烤箱继续烹制入味，选择『单点分层蒸煮』模式，时长 180 分钟。

❹ 出锅后，将料汁滗出，倒入炒锅中，加入 20g 淀粉制成的水淀粉勾芡，淋上香油 20ml 增香，将芡汁浇在鸭子上，菜品即成。

WORKING PROCESS

1. First of all, put the duck in a basin, coloring the duck by soy sauce 50ml, grill by using the Universal Steam Oven, select "Single Point Stratified Grill" mode for 5 minutes.

2. Make the sauce at the same time, pour oil 130ml into a heated wok, stir-fry with shred pork belly, add scallion 150g, sliced ginger 50g, stir-fry till fragrance. Add Sichuan preserved leaf mustard, pour sesame oil 80ml, soaked chili 20g, cooking wine 150ml, soy sauce 20ml, sugar 20g, ground pepper 5g, pour water 3L, add salt 10g after water boiled, then the sauce is done.

3. Mix the sauce with grilled duck, then steam it by using the Universal Steam Oven, select "Single Point Stratified Steam" mode for 180 minutes.

4. Separate the sauce and the duck after the dish done, pour the sauce into wok, thicken the sauce with water-starch mash, pour sesame oil 20ml to improve the flavor, pour the sauce over the duck, the dish done.

中国大锅菜

蒸烤箱卷（纪念版）

The Big-Wok-Made Cuisine of China, Food Volume of Steam Oven (Commemorative Edition)

菜品特点

特色 冬菜扒鸭是一道传统名菜，在宴会中占据重要地位，据有关人士回忆，这道菜曾经出现在开国第一宴的餐桌上。冬菜以四川南充所产为佳，誉满全国，是『四川四大腌菜』之一。

品味 冬菜具有酱香味和辛香味，香气浓郁，味道鲜美，质地脆嫩，咸淡适口。鸭肉经过长时间的炖制，已经十分软烂，肉质鲜嫩，冬菜的鲜香和里脊的肉香融入鸭肉之中，辅以香油的浓郁香味，味道奇美无比。

品相 料汁的颜色奠定了整道菜的品相，冬菜具有酱香味，加上酱油，菜品呈现酱红色，从菜色可知鸭肉十分入味，视觉效果诱人。

营养价值 四川冬菜的主要原料是芥菜，芥菜含有丰富的维生素A、维生素B族、维生素C和维生素D，具有提神醒脑、解除疲劳和通便之功效，并且芥菜的组织较粗硬，含有大量膳食纤维，具有明目和通便之功效。鸭肉中含有丰富的蛋白质，容易被人体吸收，所含B族维生素和维生素E较其他肉类多，能有效抵抗脚气病、神经炎和多种炎症，还能抗衰老。

韩酱焗排骨

Name: Baked Pork Chop with Korean Chili Sauce

制作人：孙立新　　中国烹饪大师

Made by: Lixin Sun　　A Great Master of Chinese Cuisine

主 料 Main Ingredient		
猪肋排：1500g 切段		
Pork Chop 1500g Cut		

配 料 Burdening		
嫩玉米：1000g 切段		
Green Corn 1000g Cut		
洋 葱：250g 切丝		
Onion 250g Shred		

调 料 Seasoning		
胡椒粉 Ground Pepper 3g		

盐 Salt 5g
料 酒 Cooking Wine 30ml
香 菜 Caraway 50g
牛 奶 Milk.......................200ml
白 糖 Sugar......................10g
韩式辣酱 Korean Chili Sauce ..700g
孜 然 Cumin.......................5g
料 油 Spicing Oil 20ml
姜 片 Minced Ginger 30g
葱 片 Sliced Scallion.........130g

备 注 Tips

1. 腌制排骨时可以用力抓揉，也是为了使汁液与排骨充分接触。

2. 韩式辣酱可以加些辣椒酱，味道更佳。

1. Stir fully the pork chop and Korean chili sauce during the marinating to make sure the sauce can be well absorbed by pork chop.

2. Add more chili to adjust the taste.

中国大锅菜

蒸烤箱卷（纪念版）

The Big-Wok-Made Cuisine of China, Food Volume of Steam Oven（Commemorative Edition）

制作方法

❶ 首先将排骨放入盆中腌制，加胡椒粉3g，盐5g，料酒30ml，葱片、姜片各30g，将葱姜使劲抓挤，使其汁液与排骨充分接触，搅拌均匀，腌制15分钟即可。

❷ 烤盘刷底油，铺上洋葱丝和大葱片100g，香菜50g垫底，然后摆入腌制好的排骨，放入万能蒸烤箱烤熟。选择『单点分层炙烤』，时长13分钟。

❸ 深烤盘加水，将嫩玉米放入水中，加牛奶200ml，白糖10g，放入万能蒸烤箱蒸熟，选择『单点分层蒸煮』模式，时长8分钟。

❹ 将韩式辣酱倒入盆中，加入孜然5g拌匀。将烤熟的排骨入盆与韩式辣酱拌匀，码入刷底油的烤盘，放入万能蒸烤箱烤制，将韩式辣酱焗在排骨上，选择『单点分层炙烤』模式，时长3分钟。出锅后摆入煮熟的玉米，菜品即成。

WORKING PROCESS

1. Marinate the pork chop at first, add ground pepper 3g, salt 5g, cooking wine 30ml, sliced scallion 30g, sliced ginger 30g, stir well and marinate for 15 minutes.

2. Brush oil at the bottom of the baking pan, place shred onion 100g and potato 100g, caraway 50g, then place marinated pork chop, grill by using the Universal Steam Oven, select "Single Point Stratified Grill" mode for 13 minutes.

3. Pour water into baking pan where place green corn, milk 200ml, sugar 10g, steam by using the Universal Steam Oven, select "Single Point Stratified Steam" mode for 8 minutes.

4. Put the Korean chili sauce into the basin, add cumin 5g and well stir. Stir the baked pork chop with Korean chili sauce, brush oil at the bottom of the baking pan and place the pork chop, bake by using the Universal Steam Oven, select "Single Point Stratified Grill" mode for 3 minutes, then add steamed corn, the dish is done.

中国大锅菜

菜品名称·韩酱焗排骨

Name: Baked Pork Chop with Korean Chili Sauce

菜品特点

特色 这是一道韩式风味的菜品，两次用辣酱腌制入味，采用烤制的方式将辣酱的风味焗在排骨上，制作方法简单，但是对火候要求较高，否则排骨会较老，不易入味。

品味 韩式辣酱不同于我国传统辣酱，由于腌制原料中加入苹果，故辣中有甜，微有果香，焗在排骨上，使排骨本身具有浓郁的香味之外，又有甜辣之味。排骨经过烤制，外焦里嫩，肉质汁水丰富，令人欲罢不能。嫩玉米加牛奶烹煮，香甜味被完美发挥出来，是排骨的上佳搭配。

品相 被韩式辣酱焗过的排骨呈现酱红之色，外表微焦，金黄的玉米十分夺目，整道菜菜色诱人，光彩夺目。

营养价值 猪排骨除含蛋白质、脂肪、维生素外，还含有大量磷酸钙、骨胶原、骨粘蛋白等，可为幼儿和老人提供钙质。玉米中的维生素含量非常高，为稻米、小麦的5到10倍，所含的丰富的植物纤维素具有刺激胃肠蠕动，抑制脂肪吸收，降低血脂水平，预防和改善冠心病、肥胖、胆结石症的发生。

红烩牛肉

Name: Stewed Beef with Red Wine Sauce

制作人：孙立新　　中国烹饪大师

Made by: Lixin Sun　　A Great Master of Chinese Cuisine

主 料　Main Ingredient	红葡萄酒 Red Wine 200ml	白 糖 Sugar........................15g
牛 肉：1500g 切块	番茄沙司 Ketchup 100g	胡椒粉 Ground Pepper 3g
Beef　1500g　Pieced	洋葱丝 Shred Onion..............30g	盐 Salt 10g
	面 粉 Powder 30g	姜 片 Sliced Ginger 30g
调 料　Seasoning	胡萝卜 Carrot 80g	葱 片 Sliced Scallion..........30g
料 酒 Cooking Wine 15ml	香芹丁 Chopped Parsley 80g	
黄 油 Butter 150ml	洋 葱 Onion......................130g	

中国大锅菜

菜品名称·红烩牛肉
Name: Stewed Beef with Red Wine Sauce

制作方法

① 首先将牛肉入盆上底味，加葱片、姜片各30g，将葱姜使劲抓挤，使其汁液与排骨充分接触，倒入料酒15ml，红葡萄酒50ml，放入香芹丁30g，胡萝卜片30g，洋葱丝30g，搅拌均匀即可。

② 烤盘刷油，倒入上好底味的牛肉及调味品，放入万能蒸烤箱烤制，选择『单点分层煎烤』模式，时长5分钟。

③ 然后制作面捞，待锅热后，放入黄油100g，将黄油炒化，加入面粉30g，将面粉炒透，火不能太大，否则易糊锅，面糊的黄色开始加深，面捞即制作完成。

④ 接着烧制红烩汁，锅热放入黄油50g，黄油化开后倒入番茄沙司100g，加入面捞，加香芹丁50g，萝卜片50g，大片洋葱100g，煸炒出香味，倒入水2L，大火烧开，加白糖15g，胡椒粉3g，盐10g，倒入红酒150ml，开锅后熬制一会儿，然后将调味品捞出，只留下红烩汁即可。

⑤ 将红烩汁倒入烤制好的牛肉，放入万能蒸烤箱烹制入味，选择『单点分层煎烤-1号色』模式，时长90分钟，菜品即成。

WORKING PROCESS

1. Marinate the beef in a basin. Add sliced scallion 30g, sliced ginger 30g, well stir to balance the sauce with beef. Pour cooking wine 15ml, red wine 50ml, chopped parsley 30g, sliced carrot 30g, shred onion 30g, stir evenly.

2. Brush oil at the bottom of the baking pan, put into the marinated beef and other seasoning, bake by using the Universal Steam Oven, select "Single Point Stratified Bake" mode for 5 minutes.

3. Then make the roux, put butter 100g into a heated wok, add powder 30g while the butter melting, stir-fry with light fire, continue stir-frying till the roux done.

4. Make the sauce. Put butter 50g into a heated wok, pour ketchup 100g while it melting, add roux, chopped parsley 50g, sliced carrot 50g, sliced onion 100g, stir-fry till fragrance, pour water 2L, fire till water boiled, add sugar 15g, ground pepper 3g, salt 10g, red wine 150ml, boil for a while and filter the seasoning, the red wine sauce is done.

5. Pour the red wine sauce on the beef, bake by using the Universal Steam Oven, select "Single Point Stratified Grill-Color No. 1" mode for 90 minutes, the dish is done.

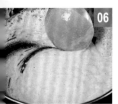

中国大锅菜

蒸烤箱卷（纪念版）

The Big-Wok-Made Cuisine of China, Food Volume of Steam Oven（Commemorative Edition）

菜品特点

特色 红烩牛肉是用地道西餐烹饪方法制作而成，是一道经典的法式大餐菜式，是学习西餐烹饪的入门菜品之一。这道菜采用慢火烹调，要求酱汁浓稠，需要制作面捞，面捞的作用相当于中餐中的芡汁，是用黄油与面粉制作而成，带有浓郁的香味。

品味 红烩汁酸中微甜，黄油的香味使得酱汁味道十分厚重，整道菜品调味料十分丰富，虽然制作好红烩汁后将调味品都捞出，但它们的味道都留在料汁中，一起衬托出牛肉的软烂醇香。番茄与牛肉堪称绝配，既能在口味上互补，又使得整道菜更富营养价值。

品相 酱汁浓稠，每一块牛肉上都挂满了红烩汁，鲜香软嫩，菜品呈现橙红色，色味俱佳。

营养价值 牛肉能提高机体抗病能力，还有暖胃作用；番茄则是含番茄红素最多的食物，有防癌功效。二者搭配不仅可发挥自身优势，更重要的是能增强补血功效。牛肉含铁较丰富，遇到番茄后，可以使牛肉中的铁更好地被人体吸收，有效预防缺铁性贫血。而在炖牛肉时，加上些番茄，能让牛肉更快变烂，更适合中老年朋友食用。

菜品名称

芝麻盐焗鸡

Name: Baked Chicken in Salt and Sesame

制作人：孙立新　　中国烹饪大师

Made by: Lixin Sun　　A Great Master of Chinese Cuisine

主　料　Main Ingredient

整　鸡：1750g　整鸡
Chicken　1750g　The original

配　料　Burdening

芝　麻：50g
Sesame　50g

调　料　Seasoning

五香粉　Five Spice Powder 5g
砂姜粉　Sand Ginger Powder...5g
海鲜酱　Seafood Paste 50g

柱候酱　Scallop Paste............40g
胡萝卜丁　Chopped Carrot ..100g
芹菜丁　Chopped Celery......100g
洋葱丝　Shred Onion............100g
香菜段　Caraway 100g
酱　油　Soy Sauce 10ml
砂姜片　Sliced Sand Ginger ...30g
南姜片　Sliced South Ginger ..30g

备　注　Tips

1. 在条件允许的情况下，鸡最好提前一天晚上就腌制好，通过更长的腌制时间，可以减少调味料的使用。

2. 此菜整鸡更适合宴会用，团餐可以将鸡腿切成块烹制。

1. when conditions permit, the chicken's better be marinated 1 night in advance which helps the chicken absorbs the seasoning and reduce the cost of the seasoning as well.

2. The whole chicken is suitable for banquet and the sliced chicken is suitable for group meal.

制作方法

❶ 将整鸡入盆腌制，撒上五香粉5g、砂姜粉5g，抹匀。加入砂姜片和南姜片各30g，海鲜酱50g，柱候酱40g拌匀，倒入胡萝卜丁100g、芹菜丁100g、洋葱丝100g、香菜段100g，使劲抓揉，挤出更多的蔬菜汁液给鸡肉上味，腌制2个小时。

❷ 将腌制好的鸡块带调料放入万能蒸烤箱蒸制成熟，选择『单点分层蒸煮』模式，时长25分钟。

❸ 出锅后将调味品去掉，刷酱油10ml上色，倒入万能蒸烤箱烤熟，选择『单点分层炙烤』模式，时长5分钟，然后将芝麻均匀撒在鸡肉上，放入万能蒸烤箱继续烤制2分钟，芝麻变成金黄色，菜品即成。

WORKING PROCESS

1. Marinate the chicken in a basin. Sprinkle five spice powder 5g, spead ginger powder 5g, sliced ginger 30g, sliced south ginger 30g, seafood paste 50g, scallop paste 40g, well stir and add chopped carrot 100g, chopped celery 100g, shred onion 100g, caraway 100g, knead and squeeze out the juice. Marinate for 2 hours.

2. Steam the marinated chicken by using the Universal Steam Oven, select "Single Point Stratified Steam" mode for 25 minutes.

3. Get rid of the seasonings afterwards, brush soy sauce 10ml to color the chicken, bake by using the Universal Steam Oven, select "Single Point Stratified Grill" mode for 5 minutes, sprinkle sesame on the chicken, continue baking by using the Universal Steam Oven for 2 minutes till the color of sesame turns into golden yellow. The dish is done.

中国大锅菜

菜品名称·芝麻盐焗鸡
Name: Baked Chicken in Salt and Sesame

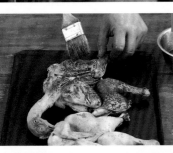

菜品特点

特色 这是一道富有东南亚风情的菜肴，砂姜和南姜是两道重要调味料，均产于东南亚，在我国华南沿海也有种植。此外，这道菜还使用了海鲜酱和柱候酱，其制作方法与客家人制作盐焗鸡类似。

品味 这道菜使用的调味品种类很多，味道十分丰富。多种香料混合在一起，产生一种诱人的异香，鸡肉外焦里嫩，经过长时间的腌制，十分入味，表面的芝麻则让香味更上一层楼。

品相 腌制好的整鸡蒸熟后呈黄白色，刷上酱油上色，炙烤出炉后，则是深深的酱红色，接近粤菜「烤乳鸽」的颜色，表面点缀芝麻，色香俱佳。

营养价值 鸡肉是蛋白质最高的肉类之一，是属于高蛋白低脂肪的食品，且易被人体吸收利用。此外，鸡肉还含有脂肪、钙、磷、铁、镁、钾、钠、维生素等营养成分。

菜品名称

红烧带鱼

Name: Braised Hairtail with Soy Sauce

制作人：王朝辉　　中国烹饪大师

Made by: Zhaohui Wang　　A Great Master of Chinese Cuisine

主　料 Main Ingredient

带　鱼：1500g　切段
Hairtails　1500g　Cut

配　料 Burdening

香　菇：250g　改朵
Xianggu Mushroom　250g　Cut

调　料 Seasoning

清　油 Oil 100ml
盐 Salt 30g
料　酒 Cooking Wine 70ml
八　角 Anise 15g
花　椒 Chinese Prickly Ash 5g
桂　皮 Cassia 10g

老　抽 Dark Soy Sauce 20ml
干辣椒 Dried Chili 20g
白　糖 Sugar 20g
醋 Vinegar 30ml
葱　段 Scallion 60g
姜　片 Sliced Ginger 40g

中国大锅菜

菜品名称 · 红烧带鱼
Name: Braised Hairtail with Soy Sauce

制作方法

① 首先腌制带鱼，加入姜片20g，葱段30g，八角10g，花椒5g，盐10g，料酒20ml，搅拌均匀，腌制15分钟。

② 烤盘刷油，将腌制好的带鱼摆入烤盘，表面刷一层明油，放入万能蒸烤箱炸制，选择『单点分层炙烤』模式，时长7分钟。

③ 烧制料汁，锅热下油100ml，加葱段30g，姜片20g，八角5g，桂皮10g，炒香后倒入料酒50ml，老抽20ml，干辣椒10g，倒入水2L，锅开后加盐20g，白糖20g，醋30ml，料汁汁即成。

④ 将炸好的鱼倒入深布菲盒内，加入香菇，倒入料汁，放入万能蒸烤箱炖熟，选择『鱼类—闷炖』模式，时长10分钟。

WORKING PROCESS

1. Marinate the hairtail in advance. Add sliced ginger 20g, scallion 30g, anise 10g, Chinese prickle ash 5g, salt 10g, cooking wine 20ml, stir well and marinate for 15 minutes.

2. Brush oil at the baking pan, place the marinate hairtail, brush another layer of oil on the hairtail, deep-fry by using the Universal Steam Oven, select "Single Point Stratified Grill" mode for 7 minutes.

3. Make the sauce. Pour oil 100ml into a heated wok, add scallion 30g, sliced ginger 20g, anise 5g, cassia 10g, stir-fry and pour cooking wine 50ml, dark soy sauce 20ml, dried chili 10g, water 2L, then add salt 20g, sugar 20g, vinegar 30ml, the sauce is done.

4. Put the deep-fried hairtail into a buffet pot, add Xianggu mushroom, pull the sauce, braise by using the Universal Steam Oven, select "Fish-Braise" mode for 10 minutes.

中国大锅菜

蒸烤箱卷（纪念版）

The Big-Wok-Made Cuisine of China, Food Volume of Steam Oven（Commemorative Edition）

菜品特点

特色 带鱼是我国东部沿海地区常见鱼类，以舟山群岛地区所产最为出名，它是深海鱼类，无法人工养殖，其鲜美较一般养殖的海产品更胜一筹。红烧为烹制带鱼最经典的方法，既保留了带鱼的鲜味，又能获得更丰富的味觉体验。

品味 带鱼以中间一根主刺为主，食用方便，肉质白嫩，异常鲜香。红烧法则能有效去除带鱼的腥味，酱汁咸中微甜，更加突出了鱼肉的鲜美。

品相 带鱼肉质雪白，可见肉质的鲜嫩，由色及味。

营养价值 带鱼富含人体必需的多种矿物元素以及多种维生素，实为老人、儿童、孕产妇的理想滋补食品，尤其适宜气短乏力、久病体虚、血虚头晕、食少羸瘦、营养不良以及皮肤干燥者食用。

菜品名称

如意菜卷

Name: Ruyi Vegetable Rolls

制作人：王朝辉　　中国烹饪大师

Made by: Zhaohui Wang　　A Great Master of Chinese Cuisine

主　料　Main Ingredient

白菜叶：1000g
Chinese Cabbage Leaves　1000g

猪肉馅：1000g
Pork Stuffing　1000g

调　料　Seasoning

清　油　Oil.........................30ml
盐　Salt 15g
胡椒粉　Ground Pepper 5g
荸　荠　Chopped Water
　　　　Chestnut.....................200g
料　酒　Cooking Wine 30ml

鸡　蛋　Egg.................2个（pcs）
白　糖　Sugar.........................20g
淀　粉　Starch........................20g
香　油　Sesame Oil 10ml
葱姜水　Scallion-Ginger
　　　　Water.....................20ml
葱　花　Chopped Scallion 20g

中国大锅菜

蒸烤箱卷（纪念版）

The Big-Wok-Made Cuisine of China, Food Volume of Steam Oven（Commemorative Edition）

制作方法

① 调制肉馅，加盐 10g，胡椒粉 5g，荸荠碎 200g，料酒 20ml，打入 2 个鸡蛋拌匀，打入葱姜水 20ml，淋上香油 10ml。

② 白菜叶放入万能蒸烤箱飞水，选择『单点分层蒸煮』模式，时长 2 分钟。

③ 飞水后的白菜叶拍上少许淀粉，加入肉馅 100g 卷好，摆入烤盘，放入万能蒸烤箱蒸熟，选择『单点分层蒸煮』模式，时长 10 分钟。

④ 烧制料汁，锅热下油 30ml，加葱花 20g 爆香，倒入热水 1L，锅开后加盐 5g，白糖 20g，料酒 10ml，倒入 20g 淀粉制成的水淀粉勾成玻璃芡。

⑤ 菜卷蒸熟后，浇上料汁，菜品即成。

WORKING PROCESS

1. Make the meat stuffing, add salt 10g, ground pepper 5g, chopped water chestnut 200g, cooking wine 20ml, 2 eggs and well stir, add scallion-ginger water 20ml, pour sesame oil 10ml.

2. Quick steam the Chinese cabbage leaves by using the Universal Steam Oven, select "Single Point Stratified Steam" mode for 2 minutes.

3. Coat starch on the cabbage leaves, stuff meat stuffing 100g and wind a roll. Place them into a baking pan, steam by using the Universal Steam Oven, select "Single Point Stratified Steam" mode for 10 minutes.

4. Make the sauce, pour oil 30ml into a heated wok, then add chopped scallion 20g, pour hot water 1L, then add salt 5g, sugar 20g, cooking wine 10ml, add water-starch mash 20g to thicken the soup.

5. Pour the sauce on the steamed vegetable rolls, the dish is done.

中国大锅菜

菜品名称·如意菜卷
Name: Ruyi Vegetable Rolls

菜 品 特 点

特色 这是一道手工菜，与饺子类似，不过面皮换成了白菜叶，可以卷入更多的馅料，满足食客口味。

品味 荸荠在肉馅中起到了关键作用，给软嫩的肉馅增添了脆爽的口感，淡淡的微甜又能突出肉馅的鲜美。调制肉馅的调料去除了猪肉的异味，两个鸡蛋带来了更滑嫩的口感，成就了馅料的美味。

品相 卷入白菜，使菜品的格调一下子就提高很多，白菜经过蒸制后变得微微有些透明，透过菜叶可以看到里面的肉馅，对喜食肉馅人士具有很大的诱惑。

营养价值 猪肉为人类提供优质蛋白质和必需的脂肪酸，可提供血红素（有机铁）和促进铁吸收的半胱氨酸，能改善缺铁性贫血。白菜营养价值丰富，是我国居民冬季餐饮中最主要的蔬菜之一。白菜含有多种营养物质，是人体生理活动所必需的维生素、无机盐及食用纤维素的重要来源，并含有丰富的钙，是预防癌症、糖尿病和肥胖症的健康食品。

菜品名称

五仁鸭子

Name: Deep-fried Duck Stuffed by Five Kinds of Kernel

制作人：王朝辉　　中国烹饪大师

Made by: Zhaohui Wang　　A Great Master of Chinese Cuisine

主　料 Main Ingredient
整　鸭: 2000g
The Whole Duck　2000g

配　料 Burdening
肉馅: 1000g
Meat Stuffing　1000g
五　仁: 750g
Five Kind of Kernel　750g

调　料 Seasoning
盐 Salt 30g
胡椒粉 Ground Pepper 5g
荸荠 Chopped Water
　　　Chestnut 200g
料酒 Cooking Wine 40ml

鸡　蛋 Egg ....2 个 (pcs)
葱姜水 Scallion-Ginger
　　　Water 20ml
香　油 Sesame Oil 20ml
酱　油 Soy Sauce 10ml
香　叶 Myrcia 3g
葱　段 Scallion 30g
姜　块 Ginger.......................30g

中国大锅菜

菜品名称·五仁鸭子

Name: Deep-fried Duck Stuffed by Five Kinds of Kernel

制作方法

① 将整鸭沿背脊片开，洗净放入盆中，加入盐20g，料酒20ml，酱油10ml，香油10ml，香叶3g，大块葱姜各30g，拌匀腌制6小时，放入万能蒸烤箱蒸熟，选择「单点分层蒸煮」模式，时长90分钟。

② 将肉馅加盐10g，胡椒粉5g，荸荠碎200g，料酒20ml，打入2个鸡蛋拌匀，打入葱姜水20ml，淋上香油10ml。

③ 将蒸熟后的鸭子拆骨铺盘，不要破坏整形，撒上一层薄淀粉，铺上搅拌好的肉馅，肉馅要抹平，盖住鸭肉，再铺上五仁，使其盖住肉馅，完成好造型后放入万能蒸烤箱烹制成熟，先选择「单点分层煎烤」模式，时长7分钟，使肉馅成熟。再选择「单点分层炙烤」模式，时长3分钟，这是为了使菜品颜色更深，口感更加酥脆。

WORKING PROCESS

1. Cut the whole duck from its back. Clean the duck and put it into a basin. Add salt 20g, cooking wine 20ml, soy sauce 10ml, sesame oil 10ml, myrcia 3g, scallion 30g, ginger 30g, marinate for 6 hours. Steam the duck by using the Universal Steam Oven, select "Single Point Stratified Steam" mode for 90 minutes.

2. Add salt 10g into meat stuffing, ground pepper 5g, chopped water chestnut 200g, cooking wine 20ml, 2 eggs, scallion-ginger water 20ml, pour sesame oil 10ml.

3. remove the bone of the duck and place into a baking pan, keep the original shape of the duck, sprinkle a layer of starch, pave the meat stuffing on it, floating. Sprinkle five kernel afterwards and cover the meat stuffing. Cook the meat stuffing by using the Universal Steam Oven, select "Single Point Stratified Grill" mode for 7 minutes. Then select "Single Point Stratified Grill" mode for 3 minutes to till the color gets deep and the duck is crispy.

菜品特点

特色 五仁鸭子是一道创新菜，将制作月饼的五仁同鸭肉、猪肉馅相结合，成为一道很有创意的菜肴。

品味 鸭肉和肉馅皆十分软嫩，其中的荸荠清脆爽口，起到了很好的调剂作用。五仁经过烤制后发出诱人的干香味，每一口咬下去，都能感受到菜品中味道的丰富，每一种味道都是十分浓厚，让人欲罢不能。

品相 整道菜品按食材分为三层，层次分明，可以切成条状摆盘，便于食用。五仁铺在最上面，经过炙烤后颜色略深，并微微烤出少许油脂，十分诱人。

营养价值 鸭肉性凉，炙烤的方法可以起到中和作用，五仁均属于坚果，富含不饱和脂肪酸，油脂较多，利于吸收，具有很高的营养价值。猪肉为人类提供优质蛋白质和必需的脂肪酸，可提供血红素（有机铁）和促进铁吸收的半胱氨酸，能改善缺铁性贫血。

香辣仔鸡

Name: Deep-fried Spring Chicken With Chili

制作人：王朝辉　　中国烹饪大师

Made by: Zhaohui Wang　　A Great Master of Chinese Cuisine

主 料 Main Ingredient
当年仔鸡：2500g　切块
Virgin Chicken　2500g　Cut into
pieces

调 料 Seasoning
清 油 Oil......................100ml

盐 Salt 20g
料 酒 Cooking Wine 20ml
淀 粉 Starch......................30g
花 椒 Chinese Prickle Ash ..10g
干辣椒 Dried Chili 50g
葱 末 Minced Scallion.........20g
姜 末 Minced Ginger 20g

备 注 Tips
如果想做得更辣可以在烧制料汁时
加入鲜辣椒。
The fresh chili can be added in the
food if you want to have stronger
taste in spicy.

中国大锅菜

The Big-Wok-Made Cuisine of China, Food Volume of Steam Oven（Commemorative Edition）

蒸烤箱卷（纪念版）

制作方法

① 将仔鸡去头、脚，剁成小块，洗净并控干水分，加入盐20g，料酒20ml，腌制15分钟。

② 烤盘刷底油，将腌制好的鸡块加入淀粉30g，拌入少许清油后放入烤盘，刷一层明油，放入万能蒸烤箱炸制，选择『单点分层炙烤』模式，时长7分钟。

③ 烧制料汁，锅热下油100ml，放入花椒10g、干辣椒50g炸香，加入葱、姜末各20g爆香，浇上料汁即成。

④ 将料汁趁热拌入炸熟的鸡块，搅拌均匀，即可盛盘。

WORKING PROCESS

1. Cut the head and feet of the chicken into small pieces, clean and dry it. Add salt 20g, cooking wine 20ml, marinate for 15 minutes.

2. Brush oil in the baking pan, sprinkle starch 30g on marinated chicken with a little oil, then brush another layer of oil, deep-fry by using the Universal Steam Oven, select "Single Point Stratified Grill" mode for 7 minutes.

3. Make the sauce. Pour oil 100ml into a heated wok, add Chinese prickle ash 10g, dried chili 50g, stir-fry till fragrance out, then add Minced scallion 20g, minced ginger 20g, the sauce is done.

4. Mix the sauce with deep-fried chicken, well stir and the dish is done.

中国大锅菜

菜品名称·香辣仔鸡
Name: Deep-fried Spring Chicken With Chili

菜品特点

特色 这是一道川渝名菜，家家会做，选用当年仔鸡，鸡肉肉质细嫩，调料中突出辣椒和花椒，给菜品赋予了浓厚的麻辣风格，辣中有香，是佐酒下饭的佳肴。

品味 这道菜为麻辣味，调味方法为双味复合型，突出了川菜麻辣鲜香的特点。鸡肉经过腌制十分入味，挂糊炸熟，外焦里嫩，干香之中保留了鸡肉特有的鲜美，与麻辣味的调味汁结合，让人直呼过瘾。

品相 火红的辣椒是这道菜最明显的标志，激发人的食欲。鸡肉炸至焦脆，表面很干，更易吸味道浓厚的料汁。

营养价值 鸡肉肉质细嫩，滋味鲜美，并富有营养，有滋补养身的作用。鸡肉中蛋白质的含量比例很高，而且消化率高，很容易被人体吸收利用，有增强体力、强壮身体的作用。中医认为鸡肉性平、温、味甘，入脾、胃经，可益气、补精、添髓。

山药木耳炒南瓜

虾皮菠菜

菜品名称

醋 椒 鱼

Name: Vinegar and Pepper Weever

制作人：王海东　　中国烹饪大师

Made by: Haidong Wang　　A Great Master of Chinese Cuisine

| **主 料 Main Ingredient**
鲈 鱼：1000g 切大片
Weever　1000g　Sliced

配 料 Burdening
洋 葱：500g 切丝
Onion　500g　Shredded | **调 料 Seasoning**
清 油 Oil..........................10ml
盐 Salt 13g
香 菜 Caraway 30g
胡椒粉 Ground Pepper 5g
料 酒 Cooking Wine 45ml
米 醋 Rice Vinegar............70ml | 白 糖 Sugar.........................3g
香 油 Sesame Oil 20ml
葱 Scallion 40g
姜 Ginger..............................40g |

中国大锅菜

菜品名称·醋椒鱼

Name: Vinegar and Pepper Weever

制作方法

❶ 将每条鲈鱼沿主刺两边片成两片，将鱼片入盆腌制，加盐8g，料酒20ml，葱段、姜片各20g，白糖3g，拌匀后腌制15分钟。

❷ 烤盘内刷底油，将腌制好的鱼片摆入烤盘，鱼上刷一层明油，放入万能蒸烤箱炸制，选择『单点分层炙烤－4号色』，时长8分钟。

❸ 炸鱼时烧制料汁，锅内倒入水1.7L，锅开后加盐5g，胡椒粉5g，料酒25ml，倒入米醋70ml，搅拌均匀，加香油20ml，放入葱丝、姜丝各20g，烹制少许，料汁即成。

❹ 将料汁浇在炸好的鱼上，倒入万能蒸烤箱烹制入味，选择『单点分层炙烤－3号色』模式，时长3分钟，出锅后撒上香菜，菜品即成。

WORKING PROCESS

1. Divide the Weever along the main thorn into 2 pieces, marinate it in a basin, add salt 8g, cooking wine 20ml, scallion 20g, sliced ginger 20g, sugar 3g, stir evenly and marinate for 15 minutes.

2. Brush oil at the bottom of a baking pan, place the marinate fish fillets into the baking pan, brush oil again on the fish and bake by using the Universal Steam Oven, select "Single Point Stratified Grill-Color No.4" for 8 minutes.

3. Make the sauce. Pour water 1.7L into a wok, boil the water and add salt 5g, ground pepper 5g, cooking wine 25ml, rice vinegar 70ml, stir evenly and pour sesame oil 20ml, shredded scallion 20g, shredded ginger 20g, cook for a while and the sauce is done.

4. Dress the sauce on the fried fish fillets, cook it by using the Universal Steam Oven, select "Single Point Stratified Grill-Color No.3" for 3 minutes, then sprinkle caraway to finish the dish.

中国大锅菜

蒸烤箱卷（纪念版）

The Big-Wok-Made Cuisine of China, Food Volume of Steam Oven (Commemorative Edition)

菜 品 特 点

特色 醋椒鱼是山东济南传统名菜之一，是一道地道的鲁菜。山东临海，近海盛产鲈鱼，是做这道菜的上好食材。此菜在清朝时就十分闻名，后来随着大批山东菜馆和厨师入京，醋椒鱼便在北京流行，但在用料上已有所不同，北京采用鳜鱼，传至京城不久也很快受到人们的青睐，近百年来成为北京著名的特色菜之一。

品味 烹饪这道菜以米醋为佳，经过烹饪，酸味淡去，只留醋香。汤味浓郁，微带酸辣，鱼肉入口鲜美，无腥腻之感，是一道开胃、解酒的佳品。

品相 汤汁颜色由于醋的缘故，略微偏深，鱼肉雪白嫩滑，点缀以少许香菜，菜色更加鲜艳，清香诱人。

营养价值 本菜谱选用鲁菜传统的鲈鱼制作此菜，鲈鱼富含蛋白质、维生素A、B族维生素、钙、镁、锌、硒等营养元素。具有补肝肾、益脾胃、化痰止咳之效，对肝肾不足的人有很好的补益作用。

番茄面包虾

Name: Tomato Bread Shrimp

制作人：王海东　　中国烹饪大师

Made by: Haidong Wang　　A Great Master of Chinese Cuisine

主　料 **Main Ingredient**	Sop　500g　Chopped	番茄酱 Ketchup.................120g
虾　肉：500g　去虾线		料　酒 Cooking Wine 13ml
Shrimp　500g　Cleaned	**调　料 Seasoning**	淀　粉 Starch......................25g
	清　油 Oil..........................35ml	葱　末 Minced Scallion.........15g
配　料 Burdening	盐 Salt 10g	姜　末 Minced Ginger 15g
面包片：500g　切丁	白　糖 Sugar......................30g	蒜　末 Minced Garlic............15g

中国大锅菜

蒸烤箱卷（纪念版）

The Big-Wok-Made Cuisine of China, Food Volume of Steam Oven（Commemorative Edition）

制作方法

❶ 将面包丁放入万能蒸烤箱烤制去水分，选择『干烤曲奇－4号色』模式，时长5分钟。将10g淀粉制成水淀粉备用。

❷ 将虾肉入盆腌制，加料酒3ml，盐5g，搅拌均匀，加淀粉15g上浆挂糊，浇上清油5ml拌匀，腌制15分钟。

❸ 烤盘刷底油，将腌制好的虾肉放入万能蒸烤箱烹制成熟，选择『单点分层炙烤』模式，时长2分30秒。

❹ 炒制料汁，锅热下油30ml，加葱、姜、蒜各15g爆香，倒入番茄酱120g翻炒，倒入水100ml，加糖30g，盐5g，料酒10ml，倒入水淀粉勾芡，浇上料汁即成。

❺ 将烹制成熟的虾仁放入料汁，然后均匀浇在面包块上，菜品即成。

WORKING PROCESS

1. Bake the sop by using the Universal Steam Oven, select "Dried Bake Cookie-Color No.4" for 5 minutes. Turn the starch 10g into water-starch mash for later use.

2. Marinate the shrimp in a basin. Pour cooking wine 3ml, salt 5g, stir evenly, coat by using starch 15g, pour oil 5ml and stir evenly, marinate for 15 minutes.

3. Brush oil at the bottom of the baking pan, place the marinated shrimp in it and cook by using the Universal Steam Oven, select "Single Point Stratified Grill" mode for 2 minutes and 30 seconds.

4. Make the sauce. Pour oil 30ml into a heated wok, add scallion 15g, ginger 15g, garlic 15g, deep-fry till fragrance. Pour ketchup 120g and stir-fry, pour water 100ml, sugar 30g, salt 5g, cooking wine 10ml, thicken the soup with water-starch mash, the sauce is done.

5. Pour the sauce on the shrimp, then sprinkle on the chopped bread. The dish is done.

中国大锅菜

菜品名称·番茄面包虾
Name: Tomato Bread Shrimp

菜品特点

特色 这是一道酸甜开胃的菜品，制作方法简单，味道香甜，非常适合用作团餐菜品。配菜除了面包，还可以选用根茎类的比较成型的蔬菜，颜色搭配鲜艳，口味上起到调剂作用。

品味 虾仁入口软嫩，肉质鲜美，以番茄酱炒制料汁，口味酸甜，更凸显了虾的美味。面包丁经过烤制，外表脆爽，入口香甜，奶香浓郁，与料汁相得益彰。

品相 料汁浓稠，红艳诱人，使得红色的虾仁更加鲜艳，面包呈金黄色，由颜色即可知酸甜宜口。

营养价值 虾肉营养价值丰富，性温，富含蛋白质，而脂肪含量较低，易于消化，是滋补的佳品。面包主要成分是面粉，富含蛋白质、碳水化合物、维生素和钙、铁、磷、钾、镁等矿物质，有养心益肾、健脾厚肠、除热止渴的功效。

蒜香凤翅

菜品名称

锅 塌 鱼

Name: Dry Fried Yellow Croaker

制作人：王海东　　中国烹饪大师

Made by: Haidong Wang　　A Great Master of Chinese Cuisine

主 料 Main Ingredient
黄 鱼：1000g　去骨去皮
Yellow Croaker　1000g
Boneless & Skinless

调 料 Seasoning
清 油 Oil...........................10ml
盐 Salt 25g
淀 粉 Starch......................50g
胡椒粉 Ground Pepper..........3g
料 酒 Cooking Wine 25ml

鸡 蛋 Egg.................3 个 (pcs)
红椒丝 Shredded Red Bell
　　　　Pepper.....................30g
葱 Scallion 80g
姜 Ginger 80g

备 注 Tips
据高家传承人透露，烹饪锅塌鱼有两
个关键：一是选好料，即用鱼必须
是当天的新鲜鱼，每次约 2 斤左右；
二是将剔骨处理好的鱼入锅烹饪时，
要掌握好火候。

According to the inheritor of the dish from Gao Family, there are two kep points of cooking Dry Fried Fish: one is the material choosing, that the fish must be fresh on the day, and about 1kg per time. The second is to control the temperature during the processing of cooking.

中国大锅菜

菜品名称 · 锅 塌 鱼
Name: Dry Fried Yellow Croaker

制 作 方 法

① 将鱼片入盆腌制，加盐 15g，料酒 15ml，葱、姜片各 30g，搅拌均匀，腌制 15 分钟左右。

② 将三个鸡蛋打散备用。将腌制好的鱼片放入淀粉盆中滚一圈，使鱼身上裹满干淀粉，再放入蛋液中过一下，使蛋液包裹住鱼身。

③ 烤盘刷底油，将加工好的鱼片码入盘中，放入万能蒸烤箱烤熟，选择「单点分层炙烤」模式，时长 5 分 30 秒。

④ 调制生汁，盆内加盐 10g，胡椒粉 3g，料酒 10ml，倒入开水 200ml，搅拌均匀即可。

⑤ 鱼片烤好后，在鱼身上码上葱丝 50g，姜丝 50g，红椒丝 30g，将生汁浇在烤好的鱼上，放入万能蒸烤箱蒸制入味。选择「单点分层蒸煮」模式，时长 2 分钟，菜品即成。

WORKING PROCESS

1. Marinate the sliced fish in a basin. Add salt 15g, cooking wine 15ml, sliced scallion 30g, sliced ginger 30g, stir evenly and marinate for 15 minutes.

2. Break 3 eggs for later use. Coat the marinated fish with starch and egg mash.

3. Brush oil at the bottom of baking pan, place the marinated sliced fish into the pan, bake by using the Universal Steam Oven, select "Single Point Stratified Grill" mode for 5 minutes and 30 seconds.

4. Make the sauce. Add salt 10g into basin, ground pepper 3g, cooking wine 10ml, boiled water 200ml, stir evenly.

5. After baking the fish, pave shredded scallion 50g, shredded ginger 50g, shredded red bell pepper 30g. Pour the sauce on the fish. Steam it by using the Universal Steam Oven, select "Single Point Stratified Steam" mode for 2 minutes, the dish is done.

中国大锅菜

蒸烤箱卷（纪念版）

The Big-Wok-Made Cuisine of China, Food Volume of Steam Oven (Commemorative Edition)

菜品特点

特色 塌是一种烹饪方法，把酥脆食品再入锅煎蒸回软，谓之塌，指将经过刀功处理的鲜嫩原料，经调味、拍粉，挂糊后，放入小油量的锅中，两面煎上色，再加入汤汁和调味品烧透入味，收尽汤汁。这是一道山东临沂地区的招牌菜，相传在20世纪30年代由临沂名厨高钧发明，并开办了『泰丰馆』，如今这道菜仍然在这家餐馆传承着。

品味 保证这道菜品味的关键是二次烹调，先煎炸，再加料汁烧制，将料汁烧尽，又将鱼肉的鲜美和料汁的鲜香充分发挥，是一道广为流传的上佳菜品。鱼肉和面糊将料汁全部吸收，鲜香味浓，软嫩无刺，既便于食用，又将鱼肉的鲜美和料汁的鲜香充分发挥。

品相 鱼肉经过煎炸色泽金黄，形状扁圆，收汁之后更显晶莹，宛如精心雕琢一般，见菜色而知美味。

营养价值 黄鱼含有丰富的蛋白质、矿物质和维生素，对人体有很好的补益作用，对体质虚弱和中老年人来说，食用黄鱼会收到很好的食疗效果。此外，黄鱼含有丰富的微量元素硒，能清除人体代谢产生的自由基，能延缓衰老，并对癌症有防治功效。

菜品名称

黑椒鸭胸

Name: Grilled Duck Breast with Black Pepper

制作人：王海东　　中国烹饪大师

Made by: Haidong Wang　　A Great Master of Chinese Cuisine

主 料 Main Ingredient

鸭　胸：1800g　整块
Duck Breast　1800g　The Original

配 料 Burdening

洋　葱：250g　切丝
Onion　250g　Shredded

芹　菜：250g　切条
Celery　250g　Shredded

调 料 Seasoning

清　油　Oil 10ml
盐　Salt 15g
黑椒碎　Black Pepper 50g
香　叶　Myrcia .... 2 片（pieces）
料　酒　Cooking Wine 25ml
胡椒粉　Ground Pepper 3g

备 注 Tips

鸭肉中心温度达到78℃即熟，如果想鸭肉更软嫩，也可以设置为78℃。
The duck meat can be cooked at 78℃, the temperature can be set as 78℃.

中国大锅菜

蒸烤箱卷（纪念版）

The Big-Wok-Made Cuisine of China, Food Volume of Steam Oven（Commemorative Edition）

制作方法

❶ 鸭胸入盆腌制，倒入洋葱丝和芹菜条，加香叶两片，盐15g，料酒25ml，胡椒粉3g，拌匀后腌制15分钟。

❷ 鸭胸的一面蘸满黑椒碎，码入刷好底油的烤盘中，烤制成熟，在一个鸭胸上插入探针，选择『单点分层炙烤－中心温度80℃』成熟，菜品即成。

WORKING PROCESS

1. Marinate the duck breast, mix with shredded onion and stripped celery, myrcia 2 pieces, salt 15g, cooking wine 25ml, ground pepper 3g, well stir and marinate for 15 minutes.

2. Sprinkle the ground pepper on one side of the duck breast, pave in the baking pan and bake till cooked, select "Single Point Stratified Grill-Core Temperature 80℃ ". The dish is done.

中国大锅菜

菜品名称·黑椒鸭胸
Name: Grilled Duck Breast with Black Pepper

菜品特点

特色 黑椒是西餐中常用的调味料，而鸭肉是我国人民十分喜爱的肉制品，这道菜品制作方法简单，价格便宜，十分适合团餐使用。使用万能蒸烤箱制作此菜，能通过探针测试菜品中心温度来控制火候，方便快捷。

品味 这道菜辛香浓郁，外焦里嫩，肉质鲜美。黑胡椒具有强烈、刺激的味道，在烤制烹调中广为使用，能够有效消除腥味。使用万能蒸烤箱能够控制鸭肉的中心温度，既保证了菜品的成熟，又能将肉质的软嫩控制在最佳火候。

品相 鸭肉外表略焦，覆盖一层黑胡椒，鸭肉烤制后会流出很多油脂，出品时需要盛盘，并尽快食用。

营养价值 鸭肉性凉，经过腌制和烤制，加上黑胡椒的辛辣，可以很大程度上中和鸭肉的凉性。鸭肉中含有丰富的蛋白质，容易被人体吸收，所含B族维生素和维生素E较其他肉类多，能有效抵抗脚气病、神经炎和多种炎症，还能抗衰老。

菜品名称

干炸带鱼

Name: Dry-fried Hairtail

制作人：王连生　　中国烹饪大师

Made by: Liansheng Wang　　A Great Master of Chinese Cuisine

主　料　Main Ingredient
带鱼：3000g　切段
Hairtail　3000g　Cut into segments

调　料　Seasoning
盐　Salt 30g
淀　粉　Starch 150g
胡椒粉　Ground Pepper 5g
料　酒　Cooking Wine 30ml
料　油　Cooking Oil 50ml

大　料　Aniseed.....................10g
花　椒　Pepper.........................3g
老　抽　Dark Soy Sauce 10ml
葱　Scallion 50g
姜　Ginger 40g

备　注　Tips
用万能蒸烤箱制作此菜要刷油并高温炙烤，相比油锅炸制，可大大减少食用油的使用量，且使菜品的品质更加稳定，设定好程序后不必担心炸糊。
This dish should be brushed with oil and grilled at high temperature in the Universal Steam Oven. Compared with frying in a frying pan, Which can greatly reduce the use the amount of oil, and make sure the quality of dish more stable. After setting the procedure, we don't need to worry about the food being burnt.

中国大锅菜

菜品名称 · 干炸带鱼
Name: Dry-fried Hairtail

制 作 方 法

❶ 先腌制带鱼，将带鱼放入盆中，加入盐30g，胡椒粉5g，料酒30ml，大料10g，花椒3g，葱段50g，姜块40g，再加入老抽10ml，增色提味，腌制20分钟。腌制好后去掉多余调味品，拍淀粉150g在带鱼上。

❷ 烤盘底刷料油，将带鱼摆入烤盘，再刷一层料油，放入万能蒸烤箱进行炸制，选择『海鲜类－高温炙烤－4号色』模式，时长约为8分钟，炸至金黄色，外焦里嫩。

WORKING PROCESS

1. Pickle the hairtail, put it into a plate, add with salt 30g, ground pepper 5g, cooking wine 30ml, aniseed 10g, pepper 3g, spring onion 50g, ginger pieces 40g, add dark soy sauce 10g to increase the color and improve taste. After pickling for 20 minutes, remove the excess seasoning, and spray starch 150g on the hairtail.

2. Brush the griddle plate with the cooking oil, put the hairtail into the griddle plate, and brush with a layer of cooking oil, put it into the Universal Steam Oven for frying, select "seafood-high temperature grilling-No.4 color" mode for about 8 minutes, and fry until the hairtail becomes golden yellow, tender with a crispy crust.

中国大锅菜

蒸烤箱卷（纪念版）

The Big-Wok-Made Cuisine of China, Food Volume of Steam Oven（Commemorative Edition）

菜 品 特 点

特色 带鱼为我国『四大海产』之一，广泛分布于我国各大海域，深受广大食客喜爱。干炸带鱼是一道传统的海派名菜，将带鱼裹上面糊炸制而成，制作方法简单，营养价值丰富。

品味 鲜香可口，外焦里嫩，且只有中间一根大骨，食用方便。

品相 面糊在万能蒸烤箱的热风作用下，如同油炸一般。表面金黄，十分诱人，内里则鱼肉白嫩，咸鲜适口。

营养价值 带鱼营养价值丰富，脂肪多为不饱和脂肪酸，具有降低胆固醇的作用。中医认为它能和中开胃、暖胃补虚，还有润泽肌肤、美容功效。

菜品名称

回 锅 肉

Name: Double Cooked Pork Slices

制作人：王连生　　中国烹饪大师

Made by: Liansheng Wang　　A Great Master of Chinese Cuisine

主　料　Main Ingredient
猪五花肉：2500g　切片
Pork Belly　2500g　Sliced

配　料　Burdening
洋　葱：1000g　切片
Onion　1000g　Sliced
青、红椒：500g　切片
Green, Red
Bell Pepper　500g　Sliced

小　料　Other Seasoning
姜：30g　切末
Ginger　30g　Minced

葱：30g　切末
Scallion　30g　Minced

调　料　Seasoning
清　油　Oil.........................300ml
盐　Salt 30g
料　酒　Cooking Wine 100ml
郫县豆瓣　Pixian Chili Bean
　　　　　Paste...................300g
干豆豉　Fermented Soya
　　　　Bean........................200g
白　糖　Sugar........................80g
生　抽　Soy Sauce 30ml
老　抽　Dark Soy Sauce........5ml

备　注　Tips
1. 炒汁时不用放水，底油给足，一定要煸炒郫县豆瓣酱出红油，可出浓郁的香辣之味。
2. 条件允许的话，配料用青蒜苗切段，则更加地道。
1. When making the marinade, no water needs to be added and the oil should be sufficient in quantity. The Pixian chili bean paste has to be fried until the hot oil becomes red and releases strong, spicy fragrance.
2. If possible, cut green garlic shoots into lengths and add as an ingredient to make the dish more authentic.

中国大锅菜

制作方法

❶ 将五花肉生切成片，烤盘刷底油，五花肉摆入烤盘放入万能蒸烤箱进行干煸，选择『肉类－高温炙烤－3号色』，时长5分钟。

❷ 在干煸的同时进行炒汁。锅热后下油300ml，油热加郫县豆瓣酱，煸炒出红油。加葱末30g，姜末30g，干豆豉200g，继续煸炒。再放入料酒100ml，生抽30ml，老抽5ml，盐30g，白糖80g，开锅后稍待一会儿便可将味汁制好。

❸ 将洋葱片和青、红椒片放入煸炒后的五花肉中，加入烧制好的味汁拌匀，继续放入万能蒸烤箱煸炒。选择『肉类－高温炙烤－3号色』模式，时长6分钟左右，至五花肉打卷即可。

WORKING PROCESS

1. Cut the pork belly into slices and put them on a baking pan before sending into the Universal Steam Oven to braise. Select "Meat-High Temperature Grill-Color No.3" mode for 5 minutes.

2. While roasting, make the dressing as follows: put oil 300ml into a heated wok; after the oil becomes hot, add Pixian chili bean paste and stew until the oil is tinted red, and then add minced scallion 30g, minced ginger 30g, dry fermented soya beans 200g and continue frying; add cooking wine 100ml, light soy sauce 100ml, dark soy sauce 5ml, salt 30g, sugar 80g. Leave it boil for a few minutes, and the dressing is successfully made.

3. Put onion slices and green and red bell pepper slices into the fried pork belly, add the dressing and mix them thoroughly before placing the pan back into the Universal Steam Oven, select "Meat-High Temperature Grill-Color No.3" mode, and roast for around 6 minutes until the sliced pork curls.

中国大锅菜

菜品名称·回锅肉

Name: Double Cooked Pork Slices

菜品特点

特色 回锅肉是一道传统的川菜名菜，俗话说「入蜀不吃回锅肉，等于没有到四川」，回锅就是再次烹调，使肉片分两段煮熟。相传清末时成都有位姓凌的翰林，因官场失意退隐，潜心研究烹饪，改良菜肴制作方法，便是今天我们所吃到的回锅肉。

品味 此菜口味独特，激发食欲，五花肉肥而不腻，入口鲜香软嫩，辣味回荡在口中，食之令人难忘，是下饭之佳品。

品相 此菜油大料足，料汁用郫县豆瓣酱煸炒而成，色泽红亮，配以青红椒片，增加鲜艳之色，更能化解油腻之感。

营养价值 回锅肉主要用料为猪五花肉，含有丰富的优质蛋白质和必需的脂肪酸；青椒、洋葱都富含维生素，而产生辛辣的辣椒素能增进食欲、帮助消化、解寒祛湿。

菜品名称

鸡 里 蹦

Name: Deep Baked Chicken and Shrimp

制作人：王连生　　中国烹饪大师

Made by: Liansheng Wang　　A Great Master of Chinese Cuisine

主 料　Main Ingredient
鸡胸肉: 2500g　切丁
Chicken Breast　2500g　Pieced
虾 仁: 500g　整仁
Shrimp　500g　The Original

配 料　Burdening
黄 瓜: 500g　切丁
Cucumber　500g　Pieced
胡萝卜: 500g　切丁
Carrot　500g　Pieced

小 料　Other Seasoning
姜: 10g　切末

Ginger　10g　Minced
葱: 10g　切末
Scallion　10g　Minced

调 料　Seasoning
清 油　Oil...........................80ml
盐 Salt 50g
酱 油　Soy Sauce 30ml
淀 粉　Starch....................150g
料 酒　Cooking Wine 30ml
胡椒粉　Ground Pepper 10g
白 糖　Sugar.......................15g
鸡 蛋　Egg.................2 个 (pcs)
料 油　Spicing Oil 50ml

备 注　Tips
用万能蒸烤箱制作此菜可以节约滑油时的用油量，而配菜的加工用蒸制代替焯水，可以留住食材中的营养物质，且颜色更加鲜亮。
It will save oil if you make this dish via the Universal Steam Oven. To steam other burdening instead of blanching them will keep the nutrition of food being inside themselves in a beautiful color.

中国大锅菜

菜品名称·鸡 里 蹦

Name: Deep Baked Chicken and Shrimp

制作方法

① 鸡丁和虾仁分别腌制上浆，分别将鸡丁和虾仁倒入盆中，打入鸡蛋2个，加盐20g，酱油30ml，加入80g淀粉拌成的水淀粉，拌均匀上劲，腌制20分钟即可。

② 烤盘刷底油，将腌制好的鸡丁和虾仁倒入烤盘，放进万能蒸烤箱进行滑油，选择『禽类－高温炙烤－3号色』模式，鸡丁时长3分钟，虾仁时长1分钟。将胡萝卜丁和黄瓜丁放入万能蒸烤箱进行焯水，选择『蔬菜蒸制』模式，99℃，胡萝卜丁时长2分钟，黄瓜丁时长1分钟。

③ 在滑油、焯水的同时进行炒汁，锅热后放入80ml底油，油热放葱、姜末各10g炒香，加开水500ml，胡椒粉8g、盐30g、淀粉70g拌成的水淀粉，大火烧开，淋入50ml料油即制成味汁。

④ 将主辅料混合，倒入味汁，拌匀后放入万能蒸烤箱进行滑炒，选择『禽类－高温炙烤－3号色』模式，3分钟后即可出锅。

WORKING PROCESS

1. Marinate and starch the chopped chicken and shrimp, put the chicken and shrimp into different basins separately. For each of the basin, add 2 eggs, salt 20g, soy sauce 30ml, water-starch mash 80g, stir and mix them evenly, set aside for 20 minutes.

2. Brush the oil at the bottom of baking pan, put the chicken and shrimp into the baking pan separately, slightly bake in the Universal Steam Oven, select "Poultry-High Temperature Grill-color No. 3" mode, 3 minutes for chicken and 1 minute for shrimp. Put the chopped carrots and cucumbers into the Universal Steam Oven, select "Steam Vegetable" mode, 99℃, 2 minutes baking for carrot and 1 minute baking for cucumber.

3. The dressing should be made at the same time, pour oil 80ml into a heated wok, add minced scallion 10g and minced ginger 10g while the oil heated, stir-fried till fragrance. Then pour boiled water 500ml, ground pepper 8g, salt 30g, water-starch mash 70g, heat till boiled, sprinkle spicing oil 50ml. Then the dressing is done.

4. Mix all ingredients with dressing, stir evenly and put into the Universal Steam Oven, select "Poultry-High Temperature Grill-color No.3" mode for 3 minutes.

菜品特点

特色 这是一道北方名菜，相传在清康熙年间由保定府厨师首创，『鸡里』与『吉利』谐音，获得康熙嘉奖：『好一个鸡里蹦！鸡、虾为水陆两鲜集萃，此菜有情有景，菜名栩栩如生。』

品味 其口味也是北方特色的咸鲜口，其烹饪方法为滑炒，使鸡肉脆嫩、虾仁软滑，鲜香之味久久留于唇齿之间。

品相 此菜颜色鲜艳，鸡肉金黄，虾肉白里透红，胡萝卜与黄瓜的色彩搭配相得益彰，相比初创时，少了几分酱色，增添几分秀色。

营养价值 此菜营养丰富，鸡肉和虾肉都富含蛋白质，而脂肪含量又较低，从中医上讲，性平、温，可温中益气，是滋补的上佳菜品。

茄汁菠萝鸡球

Name: Baked Chicken with Pineapple in Ketchup

制作人：王连生　　中国烹饪大师

Made by: Liansheng Wang　　A Great Master of Chinese Cuisine

主　料 Main Ingredient
鸡胸肉：3000g　切丁
Chicken Breast　3000g　Pieced

配　料 Burdening
菠　萝：750g　切丁
Pineapple　750g　Pieced
青、红椒：250g　切片
Green and Red
Bell Pepper　250g　Sliced

小　料 Other Seasoning
蒜：10g　切末
Garlic　10g　Minced

调　料 Seasoning
清 油 Oil......................250ml
盐 Salt................................40g
料 酒 Cooking Wine.........15ml
白 糖 Sugar....................140g
胡椒粉 Ground Pepper.........10g

老 抽 Dark Soy Sauce........5ml
鸡 蛋 Egg.................2 个 (pcs)
淀 粉 Starch....................350g
白 醋 White Vinegar..........20ml
番茄汁 Ketchup................300ml
料 油 Spicing Oil 5ml

制作方法

❶ 将鸡胸肉切丁，放入盆中，加盐 30g，胡椒粉 5g，糖 20g，老抽 5ml，料酒 15ml，鸡蛋两个，将调料揉均匀腌制，加淀粉 200g，明油 100ml，上糊定型。

❷ 烤盘内刷底油，将腌制好的鸡块放入万能蒸烤箱内滑油，选择『禽类－高温炙烤』模式，时长 5 分钟。将菠萝和青红椒片分开摆入烤盘，放入万能蒸烤箱蒸制，选择『蒸制蔬菜』模式，99℃，时长两分钟。

❸ 在滑油、蒸制的过程中炒汁，锅热放底油 100ml，油热后加蒜末 10g，番茄汁 300ml，水 150ml，盐 10g，胡椒粉 5g，糖 120g，白醋 20ml，150g 淀粉制成的水淀粉，锅开后淋入 5ml 料油即制成料汁。

❹ 将加工好的主辅料倒入布菲盒中，淋入料汁拌匀，即可出锅。

WORKING PROCESS

1. Cut the chicken into pieces, put it into a basin, mix with salt 30g, ground pepper 5g, sugar 20g, dark soy sauce 5ml, 2 eggs, starch 200g, oil 100ml. Make them well stir, knead to balance the seasoning to marinate, to paste on finalize shape with starch mash.

2. The baking pan should be brushed by oil, the marinated chicken should be baked, select "Poultry-High Temperature Grill" mode for 5 minutes. Put the pineapple and red and green bell pepper into baking pan separately, steam in the Universal Steam Oven, select "Steam Vegetable" mode, 99 degree centigrade, prolongs 2 minutes.

3. During baking and steaming process, the dressing should be made at another wok. Pour oil 100ml into the heated wok, heat the oil and then put minced garlic 10g, ketchup 300g, water 150ml, salt 10g, ground pepper 5g, sugar 120g, white vinegar 20ml, water-starch mash 150g, the spicing oil will be poured into the wok after the dressing boiling.

4. Mix all these ingredients into a buffet pot, pour the dressing and the dish is done.

中国大锅菜

菜品名称 · 茄汁菠萝鸡球

Name: Baked Chicken with Pineapple in Ketchup

菜品特点

特色 此菜为广式名菜，由潮汕著名点心女状元郭丽文师傅于1998首创，虽历史不长，但自问世以来便广受欢迎，年轻食客尤为喜爱。制作时要将鸡肉挂糊定型，经过滑油，表皮香脆，而肉质软嫩。

品味 此菜以热带名果菠萝和鸡肉结合起来，酸甜可口，而菠萝中特有的菠萝胺使鸡肉更加软嫩，形成十分独特的口感。鸡肉与水果的搭配使其没有油腻之感；而水果因鸡肉味道更加丰富，令人回味无穷，百吃不厌。

品相 由于加入番茄酱，此菜色泽红亮，从红中能感觉到酸甜的气味，给人可口之感，加之搭配造型十分美观，能看出广式菜品的精致。

营养价值 鸡肉肉质细嫩，滋味鲜美，并富有营养，有滋补养身的作用。菠萝含有大量的果糖，葡萄糖，维生素A、B、C，磷，柠檬酸和蛋白酶等物，所含的蛋白质分解酵素有分解用，有增强体力、强壮身体的作用。鸡肉中蛋白质的含量比例很高，而且消化率高，很容易被人体吸收利蛋白质及助消化的功能，有助于消解油腻。番茄酱中除了番茄红素外还有B族维生素、膳食纤维、矿物质、蛋白质及天然果胶等，与新鲜番茄相比较，番茄酱里的营养成分更容易被人体吸收。

莲藕鸭片

菜品名称

香酥翅中

Name: Crispy Fried Chicken Wings

制作人：王连生　　中国烹饪大师

Made by: Liansheng Wang　　A Great Master of Chinese Cuisine

主 料 Main Ingredient

鸡翅中：4000g 原型
Chicken Wings　4000g　The Original

小 料 Other Seasoning

姜：30g 切片
Ginger　30g　Sliced

葱：30g 切条
Scallion　30g　Slit

调 料 Seasoning

清 油 Oil.............................5ml
盐 Salt................................30g
料 酒 Cooking Wine.........30ml
胡椒粉 Ground Pepper...........3g
老 抽 Dark Soy Sauce......10ml
鸡 蛋 Egg.................3 个 (pcs)
淀 粉 Starch....................100g

备 注 Tips

鸡肉出油较多，烤制时刷少许明油即可，将皮下的油脂烤出，更加有酥脆之感。

Only a little clear oil is required for roasting since chicken meat itself produces oil. Melting the fat under the skin when roasting makes it taste even more crispy.

中国大锅菜

菜品名称 · 香酥翅中
Name: Crispy Fried Chicken Wings

制作方法

❶ 将鸡翅中进行腌制，放入葱条30g，姜片30g，盐30g，料酒30ml，老抽10ml，打入3个鸡蛋，搅拌均匀腌制20分钟左右，腌制完成后拍少许淀粉100g，加入少许明油5ml。

❷ 烤盘上架不锈钢烤架，将处理完的鸡翅中在烤架上摆好进行烤制，选择『禽类－高温炙烤』模式，3号色，时长7分钟。

WORKING PROCESS

1. The Chicken wings must be pickled with scallion 30g, sliced ginger 30g, salt 30g, cooking wine 30ml, ground pepper 3g, dark soy sauce 10ml and 3 eggs. Stir and mix well for about 20 minutes. Coat with starch 100g and oil 5ml afterwards.

2. Place the baking rack on the baking pan, prepare the chicken wings on the tray, select "Poultry-High Temperature Grill" mode, color No.3, for 7 minutes.

中国大锅菜

蒸烤箱卷（纪念版）

The Big-Wok-Made Cuisine of China, Food Volume of Steam Oven (Commemorative Edition)

菜品特点

特色 香酥翅中是一道简单、味美的菜肴，将翅中加葱、姜和调味料腌制，使之入味，然后烤制即可。鸡翅为鸡肉中之上品，瘦而不柴，肉质细嫩，适合烤制食用。此菜对火候的要求比较高，控制得当方可出香酥之感，过则焦煳。

品味 此菜应『香酥』之名，鸡皮酥脆，鸡肉软嫩。

品相 由于腌制时添加了少许酱油，故颜色较深，经过烤制，鸡肉由黄变为金黄，再变为深黄，由色便觉酥脆之感。

营养价值 鸡肉性平、温、味甘，入脾、胃经，可益气、补精、添髓。而鸡翅中的胶原蛋白含量更加丰富，对于保持皮肤光泽、增强皮肤弹性均有好处。

菜品名称

豆豉蒸鲈鱼

Name: Steamed Weever with Fermented Soya Beans

制作人：王万友　　中国烹饪大师

Made by: Wanyou Wang　　A Great Master of Chinese Cuisine

主　料　Main Ingredient
鲈　鱼：3000g
Weever　3000g

配　料　Burdening
豆　豉：150g　切碎
Fermented Soya Bean　150g
Chopped
红　椒：100g　切丝
Red Bell Pepper　100g　Shredded

调　料　Seasoning
清　油　Oil......................130ml
盐　Salt 10g
蒸鱼豉油　Soy Sauce...........30ml
小苏打　Baking Soda..............3g
胡椒粉　Ground Pepper 3g
老　抽　Dark Soy Sauce......15ml
料　酒　Cooking Wine 20ml
鸡　蛋　Egg.................2 个 (pcs)
豆　豉　Fermented Soya
　　　　Bean......................150g
生　粉　Powder....................40g
姜　末　Minced Ginger 10g
葱　Scallion 20g
蒜　末　Minced Garlic............20g

中国大锅菜

蒸烤箱卷（纪念版）

The Big-Wok-Made Cuisine of China, Food Volume of Steam Oven（Commemorative Edition）

制作方法

❶ 首先腌制鲈鱼，放入盆中，倒入料酒 20ml，老抽 15ml，盐 10g，胡椒粉 3g，生粉 40g，小苏打 3g，打入两个鸡蛋，抓揉上劲。再逐次加一点点水，揉进肉中，倒入清油 30ml，腌制 15 分钟即可。

❷ 然后炒制豉汁，锅热下油 100ml，加葱、姜末 10g，蒜末 20g，炒香，加豆豉 150g，多炒一会，炒出香味，倒入水 150ml，烧开即可。

❸ 将腌制好的鱼沿背脊切一刀，方便入味，摆入烤盘，浇上豉汁，放入万能蒸烤箱进行烹制，选择『单点分层蒸煮』模式，时长 10 分钟。

❹ 出锅后铺上葱丝 10g 和红椒丝 100g，锅内下油 30ml，油热后浇在鱼身上，表面淋上一层蒸鱼豉油 30ml，即可盛盘。

WORKING PROCESS

1. Marinate the weever in a basin. Pour cooking wine 20ml, dark soy sauce 15ml, salt 10g, ground pepper 3g, powder 40g, baking soda 3g, 2 eggs, knead it and pour water separately, pour oil 30ml, marinate for 15 minutes.

2. Make the fermented soya bean sauce. Pour oil 100ml into a heated wok, add minced scallion 10g, minced ginger 10g, minced garlic 20g, stir-fry till fragrance, then add fermented soya bean 150g, continue to stir-fry and pour water 150ml till boiled.

3. Cut the marinated fish along its back, it makes the fish to be salted easily, place the weever into the baking pan, dress the fermented soya bean sauce, cook by using the Universal Steam Oven, select "Single Point Stratified Steam" mode for 10 minutes.

4. Pave the shredded scallion and red bell pepper, pour oil 30ml into the wok, then pour the boiled oil on the fish, then pour a layer of soy sauce 30ml, the dish is done.

中国大锅菜

菜品名称 · 豆豉蒸鲈鱼

Name: Steamed Weever with Fermented Soya Beans

菜品特点

特色 鲈鱼肉质白嫩、清香，没有腥味，肉为蒜瓣形，最宜清蒸。豆豉在古代被称为『幽菽』，已经有两千多年的历史，是发酵豆制品调味料，鲜美可口、香气独特。制作此菜十分简单，越是简单的菜，其成败往往体现在细节上，似易非易。

品味 鲈鱼本身极为鲜美，腌制时间不宜过长，否则肉质易较硬，影响口感。豆豉香气浓郁，不同于鱼肉的鲜香，让这道菜整体的味觉更加厚重。蒸好后一定要先浇热油，再淋蒸鱼豉油，否则就如酱油泡的咸鱼，鱼肉也泄了。

品相 鲈鱼背脊处要开一刀，这样蒸后不易变形，且肉厚的地方容易熟。豆豉颜色较深，与鱼肉的嫩白形成鲜明对比，将鱼肉的鲜香更加衬托出来，淋上热油后，冒着腾腾热气，鲜美无比。

营养价值 鲈鱼富含蛋白质、维生素A、B族维生素、钙、镁、锌、硒等营养元素；具有补肝肾、益脾胃、化痰止咳之效，对肝肾不足的人有很好的补益作用。

菜品名称

柠檬蒸虾

Name: Steamed Shrimps with Lemon

制作人：王万友　　中国烹饪大师

Made by: Wanyou Wang　　A Great Master of Chinese Cuisine

主 料 Main Ingredient	Lemon 250g Sliced	盐 Salt 15g
鲜 虾：2000g 去虾线	粉 丝：100g 切段	料 酒 Cooking Wine 20ml
Shrimps 2000g Cleaned	Bean vermicelli 100g Cut	蒜 末 Minced Garlic...........20g
		葱 花 Minced Scallion..........5g
配 料 Burdening	调 料 Seasoning	
柠檬：250g 切片	清 油 Oil........................20ml	

中国大锅菜

菜品名称·柠檬蒸虾
Name: Steamed Shrimps with Lemon

制作方法

❶ 将粉丝用剪刀剪成段，泡水少许后捞出，粉丝内加葱花 5g，料酒 20ml，蒜末 10g，盐 5g，清油 20ml，拌匀后码入烤盘。

❷ 将虾洗净去虾线，摆在粉丝上，撒上蒜末 10g，盐 10g，铺上一层柠檬片。

❸ 将烤盘放入万能蒸烤箱进行烹制，选择『单点分层蒸煮』模式，时长 8 分钟。

WORKING PROCESS

1. Cut the bean vermicelli into pieces, soak it and take out, mix with minced scallion 5g, cooking wine 20ml, minced garlic 10g, salt 5g, oil 20ml, well stir and place into baking pan.

2. Clean the shrimps, place on the bean vermicelli, sprinkle minced scallion 10g, salt 10g, and pave a layer of sliced lemon.

3. Cook by using the Universal Steam Oven, select "Single Point Stratified Steam" mode for 8 minutes.

菜品特点

特色 蒸虾是一道最普通不过的菜肴，而柠檬给这道菜增添了许多清新之感，格调立刻就高了许多。团餐中可以加入粉丝，既有利于丰富味道，又能节约成本，一举两得。

品味 虾肉鲜美，直接蒸制食用已然是人间美味，加入柠檬后，口感酸甜，带有淡淡的清香。粉丝要事先泡制，加少许调料腌制，烹制中能充分吸收虾的鲜味，软嫩鲜香。

品相 鲜黄的柠檬除带来口感上的变化外，色彩给人以清新之感，与粉丝的晶莹柔软搭配相宜。

营养价值 虾类营养丰富，且肉质松软，易消化，对身体虚弱以及病后需要调养的人是极好的食物；虾中含有丰富的镁，镁对于心脏活动具有重要的调节作用，且能够很好地保护心血管系统。

红烧猪手

菜品名称

浓汤奶白菜

Name: Pottage Chinese Cabbage

制作人：王万友　　中国烹饪大师

Made by: Wanyou Wang　　A Great Master of Chinese Cuisine

主 料 Main Ingredient
奶白菜：2000g　切段
Chinese Cabbage　2000g　Cut

调 料 Seasoning
美极鸡汁 Maggi Chicken
　　　　　Sauce.................30ml
盐 Salt 15g

白 糖 Sugar.......................10g
水淀粉 Starch......................30g

中国大锅菜

蒸烤箱卷（纪念版）

The Big-Wok-Made Cuisine of China, Food Volume of Steam Oven（Commemorative Edition）

制作方法

❶ 将奶白菜摆入蒸盘，放入万能蒸烤箱飞水，选择『单点分层蒸煮』模式，时长 3 分钟。

❷ 烧制浓汤汁，锅内倒入水 2L 烧开，加盐 15g，白糖 10g，倒入美极鸡汁 30ml，搅拌均匀，倒入水淀粉 30g，勾成薄玻璃芡。

❸ 将飞水后的白菜夹入深烤盘，倒入浓汤汁，放入万能蒸烤箱烹制入味，选择『单点分层蒸煮』模式，时长 5 分钟，菜品即成。

WORKING PROCESS

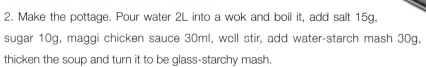

1. Place the Chinese cabbage into the baking pan, quick steam it by using the Universal Steam Oven, select "Single Point Stratified Steam" mode for 3 minutes.

2. Make the pottage. Pour water 2L into a wok and boil it, add salt 15g, sugar 10g, maggi chicken sauce 30ml, well stir, add water-starch mash 30g, thicken the soup and turn it to be glass-starchy mash.

3. Place the steamed Chinese cabbage into the baking pan, pour the pottage and cook it by using the Universal Steam Oven, select "Single Point Stratified Steam" mode for 5 minutes. Then the dish is done.

中国大锅菜

菜品名称 · 浓汤奶白菜
Name: Pottage Chinese Cabbage

菜 品 特 点

特色 这是一道汤菜，是百姓餐桌上的常客，其美味的关键就是烧制浓汤汁，不同厨师有着不同的选材。如果要烧制一锅上好的浓汤汁，这几味原材料应该是必不可少的，如鸡油、干贝、蟹肉、海米等，烧制出来的汤汁颜色黄亮，晶莹剔透。而在团餐中，可以使用美极鸡汁，既味道鲜美，又方便快捷。

品味 菜品的美味全在汤汁，一锅合格的浓汤一定要做到鲜美、浓郁，令人唇齿留香，奶白菜不要火太大，清脆爽口为佳。

品相 汤汁淡黄，由于芡汁的缘故，略有晶莹之感，浓浓的汤汁中一眼便知藏有无限的鲜美，翠绿的奶白菜更将菜品的营养与清淡彰显得淋漓尽致。

营养价值 这是一道养生菜，汤汁可与奶白菜一同品尝，营养丰富又清淡爽口。奶白菜有很好的清热去火的作用，奶白菜对于防治乳腺疾病有好处，也可以提高免疫力。同时奶白菜也可以解毒，有亮发的作用。

07

08

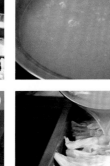

09

10

11

12

菜品名称

五花烧海带

Name: Braised Pork Belly with Kelp

制作人：王万友　　中国烹饪大师

Made by: Wanyou Wang　　A Great Master of Chinese Cuisine

主 料 Main Ingredient
五花肉：1500g 切块
Pork Belly　1500g　Cut

配 料 Burdening
海带：1000g 切丝
Kelp　1000g　Shredded

调 料 Seasoning
清 油 Oil..........................100ml
盐 Salt..................................20g
老 抽 Dark Soy Sauce......50ml
大 料 Aniseed.......................5g
桂 皮 Cassia...........................3g
料 酒 Cooking Wine.........30ml

白 糖 Sugar.....................110g
胡椒粉 Ground Pepper...........5g
姜 片 Sliced Ginger...........20g
葱 段 Scallion....................30g

中国大锅菜

菜品名称 · 五花烧海带
Name: Braised Pork Belly with Kelp

（制作方法）

① 将五花肉切成2cm见方的小块，放入万能蒸烤箱飞水，选择『单点分层蒸煮』模式，时长15分钟。

② 炒制糖色，锅热下油20ml，倒入白糖100g，将白糖炒化，呈褐色，倒入开水150ml，糖色即成，盛入碗中。

③ 烧制料汁：锅热下油80ml，加葱段30g，姜片20g，大料5g，桂皮3g，料酒30ml，倒入糖色汁，沿锅边倒入老抽50ml，加盐20g，白糖10g，胡椒粉5g，倒入开水2L，烧开即可。

④ 将海带丝放入飞水后五花肉中，倒入料汁，放入万能蒸烤箱烹制成熟，选择『单点分层蒸煮』模式，时长90分钟。

WORKING PROCESS

1. Cut the pork belly into 2cm-squared blocks, quick steam by using the Universal Steam Oven, select "Single Point Stratified Steam" mode for 15 minutes.

2. Stir-fry the caramel. Pour oil 20ml into a heated wok, add sugar 100g, stir-fry it till melting and the color turns into brown, pour boiled water 150ml, the caramel is done.

3. Make the sauce. Pour oil 80ml into a heated wok, add scallion 30g, sliced ginger 20g, aniseed 5g, cassia 3g, cooking wine 30ml, pour the caramel, pour dark soy sauce 50ml, salt 20g, sugar 10g, ground pepper 5g, boiled water 2L, boil the soup.

4. Mix the shredded kelp with steamed pork belly, dress the sauce, cook by using the Universal Steam Oven, select "Single Point Stratified Steam" mode for 90 minutes.

中国大锅菜

蒸烤箱卷（纪念版）

The Big-Wok-Made Cuisine of China, Food Volume of Steam Oven (Commemorative Edition)

菜品特点

特色 这是一道红烧菜肴，以红烧肉为基础，加入海带作为配菜，能缓解五花肉的肥腻，亦使菜品的营养价值更加丰富。

品味 五花肉的经过长时间烧制，已是肥而不腻，肥瘦层次分明，肥肉香味醇厚，瘦肉浓香适口，海带又给菜品增添了许多海的味道，肉香之外更有鲜美的品味，给厚重的口味带来清鲜。

品相 五花肉层次分明，散发着浓郁的香味，十分诱人。

营养价值 五花肉含有丰富的优质蛋白质和必需的脂肪酸，并提供血红素（有机铁）和促进铁吸收的半胱氨酸，能改善缺铁性贫血；具有补肾养血，滋阴润燥的功效。海带是一种营养价值很高的蔬菜，同时具有一定的药用价值。含有丰富的碘等矿物质元素。海带含热量低、蛋白质含量中等、矿物质丰富。研究发现，海带具有降血脂、降血糖、调节免疫、抗凝血、抗肿瘤、排铅解毒和抗氧化等多种生物功能。

菜品名称

醋熘鸡条

Name: Sweet Sour Chicken Fillets

制作人：王兆志　　中国烹饪大师

Made by: Zhaozhi Wang　　A Great Master of Chinese Cuisine

主 料 Main Ingredient	青 椒：250g 切片	酱 油 Soy Sauce 20ml
鸡腿肉：2500g 切条	Green Bell Pepper 250g Sliced	白 糖 Sugar.......................20g
Drumstick 2500g Stripped	红 椒：250g 切片	醋 Vinegar.........................30ml
	Red Bell Pepper 250g Sliced	淀 粉 Starch.......................30g
配 料 Burdening		香 油 Sesame Oil 20ml
冬 笋：250g 切片	**调 料 Seasoning**	葱 片 Sliced Scallion...........30g
Winter Bamboo Shoots 250g	清 油 Oil.......................60ml	蒜 片 Sliced Garlic.............30g
Sliced	盐 Salt 25g	姜 片 Sliced Ginger 30g

制作方法

❶ 首先腌制鸡条，加盐20g，酱油10ml，葱姜水20ml，边搅拌边加淀粉20g，打入两个鸡蛋清，搅拌均匀，腌制15分钟。

❷ 烤盘刷底油，倒入腌制好的鸡条，淋少许明油，放入万能蒸烤箱滑油。选择『单点分层炙烤』模式，时长8分钟。

❸ 烤盘刷底油，将冬笋片和青、红椒片放入烤盘，拌入盐5g，淋上明油，放入万能蒸烤箱滑油，选择『单点分层炙烤』模式，时长1分钟。

❹ 烧制料汁：锅热下油50ml，放入葱、姜、蒜片各30g，爆香后加白糖20g，酱油10ml，醋30ml，盐5g，倒入开水200ml。锅开倒入10g淀粉制成的水淀粉勾芡，加香油20ml，料汁即成。

❺ 将料汁和配菜倒入鸡条中，搅拌均匀，放入万能蒸烤箱烹制入味，选择『单点分层煎烤』模式，时长两分钟，菜品即成。

WORKING PROCESS

1. Marinate the chicken strips at first. Add salt 20g, soy sauce 10ml, scallion and ginger water 20ml, stir it and add starch 20g, 2 eggs white, stir evenly and marinate for 15 minutes.

2. Brush oil at the baking pan, place the marinated chicken strip, pour a little oil and quick bake by using the Universal Steam Oven, select "Single Point Stratified Grill" mode for 8 minutes.

3. Brush oil at the baking pan, place the sliced winter bamboo shoots, sliced red and green bell pepper, add salt 5g, pour oil, quick bake by using the Universal Steam Oven, select "Single Point Stratified Grill" mode for 1 minute.

4. Make the sauce. Pour oil 50ml into a heated wok, add sliced scallion 30g, sliced ginger 30g, sliced garlic 30g, quick deep-fry and add sugar 20g, soy sauce 10ml, vinegar 30ml, salt 5g, boiled water 200ml, then add water-starch mash 10g while the water boiled to thicken the soup. Then add sesame oil 20ml, the sauce is done.

5. Pour the sauce onto chicken strips, well stir and bake by using the Universal Steam Oven, select "Single Point Stratified Grill" mode for 2 minutes. The dish is done.

中国大锅菜

菜品名称·醋熘鸡条
Name: Sweet Sour Chicken Fillets

菜品特点

特色 熘是中国传统烹饪方法之一，进行醋熘烹调需要加入糖调整滋味，否则过于刺激。醋熘鸡条是醋熘烹调的经典菜式，使用万能蒸烤箱十分方便制作此菜，只要将料汁调好，便可十分入味，不亚于小锅炒制。

品味 这道菜酸甜可口，醋香浓郁。鸡肉的火候把握很重要，既要成熟，又要保证肉质鲜嫩。芡汁过厚则味酸，过薄则味寡，恰到好处是均匀入味。配菜鲜嫩可口，是对肉食很好的调剂。

品相 芡汁锁定了菜品的味道，不薄不厚，均匀包裹住鸡条，配以红、绿、白三色，色泽诱人，顿生食意。

营养价值 鸡肉肉质细嫩，滋味鲜美，并富有营养，有滋补养身的作用。鸡肉中蛋白质的含量比例很高，而且消化率高，很容易被人体吸收利用，有增强体力、强壮身体的作用。

菜品名称

干炸丸子

Name: Croquette

制作人：王兆志　　中国烹饪大师

Made by: Zhaozhi Wang　　**A Great Master of Chinese Cuisine**

主　料　Main Ingredient

肉　馅：2000g
Meat Stuffing　2000g

调　料　Seasoning

清　油　Oil.........................15ml
葱姜水　Ginger Water.........200ml

料　酒　Cooking Wine.........30ml
盐　Salt...................................25g
鸡　蛋　Egg................2 个 (pcs)
淀　粉　Starch.......................15g
酱　油　Soy Sauce..............20ml
椒　盐　Spiced Salt................10g

备　注　Tips

要用葱姜水调馅，馅中不可见葱姜，
因为加葱的话容易炸黑。
Use the ginger and scallion water to
make the stuffing but not directly use
ginger and scallion which is easily to
be burnt during deep-frying.

制作方法

❶ 首先调制肉馅，分次将葱姜水200ml打入肉馅，加料酒20ml，摔打上劲至肉馅较软。上劲后加调味料，盐25g，料酒10ml，打入2个鸡蛋，加淀粉15g，酱油20ml，搅拌均匀。

❷ 烤盘刷油，将肉馅攥入手中，从虎口处挤出，制成肉丸，摆入烤盘。

❸ 将肉丸上刷少许明油，放入万能蒸烤箱进行炸制，选择「单点分层炙烤」模式，时长10分钟，出锅后撒上椒盐10g，菜品即成。

WORKING PROCESS

1. Make the meat stuffing. Mix the ginger water with meat stuffing over and over again. Pour cooking wine 200ml. Knead the meat stuffing and add salt 25g, cooking wine 10ml, 2 eggs, starch 15g, soy sauce 20ml, stir evenly.

2. Brush oil on the baking pan. Clutch meat stuffing by hand, extrude stuffing and make the meatballs, place them into the baking pan.

3. Brush a little oil on the meatballs, bake by using the Universal Steam Oven, select "Single Point Stratified Grill" mode for 10 minutes. Sprinkle spiced salt 10g. The dish is done.

菜品特点

特色 干炸丸子算是京味儿菜的代表之一了，但它却是一道地道的鲁菜。鲁菜早年间来到京城时因其口感偏咸，所以有过一段儿水土不服的日子，后来经过名厨的不断改良，减轻了盐量并在其中添加了一些适合当地人口味的调料，逐渐形成了今天的味道。

品味 制作这道菜，肉馅的肥瘦比例以三七或四六为佳，这样炸出来的丸子才能油润有余，调馅时加葱姜水是关键，去腥提鲜，馅料中可以加少许黄酱调味，味道更加鲜美。丸子经过炸制，外焦脆，里软嫩，咸鲜适口，不柴亦不软，蘸椒盐而食，这种美味大概是许多老北京人美好的回忆吧。

品相 丸子色泽焦黄，宛如穿了一层焦酥的外衣。咬一口，内里肉质软嫩，仿佛饱含汁水，并且十分有质感。

营养价值 猪肉是目前人们餐桌上重要的动物性食品之一，猪脊肉含有人体生长和发育所需的丰富的优质蛋白、脂肪、维生素等，而且肉质较嫩，易消化。

菜品名称

酱烧豆腐

Name: Stewed Tofu in brown sauce

制作人：王兆志　　中国烹饪大师

Made by: Zhaozhi Wang　　A Great Master of Chinese Cuisine

主　料 Main Ingredient	料　油 Spicing Oil 30ml	香葱碎 Chopped Chive 30g
豆　腐：2500g 切块	八　角 Anise 5g	葱　片 Sliced Scallion 20g
Tofu　2500g　Cut	甜面酱 Sweet Bean Paste .. 120g	姜　片 Sliced Ginge 20g
	料　酒 Cooking Wine 20ml	蒜　片 Sliced Garlic 20g
调　料　Seasoning	盐 Salt 20g	
清　油 Oil 50ml	淀　粉 Starch 10g	

制作方法

❶ 首先将豆腐放入万能蒸烤箱焯水，选择『单点分层蒸煮』模式，时长8分钟。

❷ 焯水时烧制料汁，锅热下油50ml，加八角5g，葱、姜、蒜片各20g炒香，加入面酱120g，料酒20ml，炒匀，倒入开水1L，加盐20g，淀粉10g制成的水淀粉，出锅前淋上料油30ml，料汁即成。

❸ 焯水后，倒出烤盘中的水，均匀倒入料汁，放入万能蒸烤箱烹制入味，选择『单点分层煎烤』模式，时长3分钟，出锅后撒上香葱碎30g，菜品即成。

WORKING PROCESS

1. Quick steam tofu by using the Universal Steam Oven, select "Single Point Stratified Steam" mode for 8 minutes.

2. Make the sauce at the same time. Pour oil 50ml into a heated wok, add anise 5g, sliced scallion 20g, sliced ginger 20g, sliced garlic 20g, stir-fry till fragrance. Add sweet bean paste 120g, cooking wine 20ml, stir-fry evenly and add boiled water 1L, add salt 20g, water-starch mash 10g, sprinkle spicing oil 30ml at the last minute and the sauce is done.

3. Filter out the water of baking pan after quick steaming, dress the sauce evenly, bake by using the Universal Steam Oven, select "Single Point Stratified Bake" mode for 3 minutes. Sprinkle chopped chive 30g afterwards, the dish is done.

中国大锅菜

菜品名称·酱烧豆腐

Name: Stewed Tofu in brown sauce

菜品特点

特色 酱烧豆腐是一道家常菜，食材易得，制作方法简单，为许多人士所喜爱。

品味 酱烧，顾名思义，在烧制过程中少不了面酱。豆腐本身具有独特的豆香，充分与料汁结合后，酱香浓郁，入口软嫩，配以香葱，品味鲜香。

品相 菜品呈酱色，隐约透着豆腐的嫩白，豆腐在烹饪中会释放水分，所以料汁较稀。

营养价值 豆腐营养丰富，含有铁、钙、磷、镁等人体必需的多种微量元素，还含有糖类、植物油和丰富的优质蛋白，素有『植物肉』之美称，拥有高蛋白，低脂肪。除此之外，也具有降血压，降血脂，降胆固醇的功效。大豆蛋白属于完全蛋白质，其氨基酸组成比较好，人体所必需的氨基酸它几乎都有，并且十分容易被人体消化、吸收。

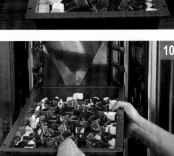

菜品名称

炸烹鱼条

Name: Deep-fried Long Li Fish Fillets

制作人：王兆志　　中国烹饪大师

Made by: Zhaozhi Wang　　A Great Master of Chinese Cuisine

主 料 Main Ingredient	**Red Bell Pepper 150g Stripped**
龙俐鱼：1500g 切条	芹 菜：150g 切条
Long Li Fish 1500g Stripped	Celery 150g Stripped
	香 菜：100g 切段
配 料 Burdening	Caraway 100g Cut
青 椒：150g 切条	
Green Bell Pepper	**调 料 Seasoning**
150g Stripped	清 油 Oil........................40ml
红 椒：150g 切条	盐 Salt 20g

葱姜水 Ginger Water...........30ml
酱 油 Soy Sauce 25ml
鸡 蛋 Egg................2 个 (pcs)
淀 粉 Starch......................35g
料 油 Spicing Oil 15ml
胡椒粉 Ground Pepper...........3g
葱 片 Sliced Scallion..........20g
姜 片 Sliced Ginger 20g
蒜 片 Sliced Garlic..............20g

中国大锅菜

菜品名称·炸烹鱼条
Name: Deep-fried Long Li Fish Fillets

制作方法

① 首先给鱼条腌制入味并上浆，加盐10g，葱姜水30ml，酱油10ml，淀粉10g，打入两个鸡蛋，再加10g淀粉，搅拌均匀，拌入料油15ml。

② 烤盘刷底油，将鱼条摆入烤盘，放入万能蒸烤箱炸制，选择『鱼类—高温炙烤』模式，时长5分钟。

③ 将青椒、红椒和芹菜倒入烤盘，撒入盐5g，淋上明油10ml，放入万能蒸烤箱滑油，选择『单点分层炙烤』模式，时长两分钟。

④ 烧制料汁，锅热下油30ml，加葱、姜、蒜片各20g炒香，加酱油15ml，倒入开水1L，锅开后加盐5g，胡椒粉3g，倒入15g淀粉制成的水淀粉勾芡，料汁即成。

⑤ 将蔬菜拌入鱼条烤盘，倒入料汁，搅拌均匀，撒上香菜段，菜品即成。

WORKING PROCESS

1. Marinate and coat fish fillets. Add salt 10g, scallion and ginger water 30ml, soy sauce 10ml, starch 10g, add in 2 eggs, then add starch 10g again, well stir and pour spicing oil 15ml.

2. Brush oil at the bottom of baking pan, place fish fillets on the baking pan, deep-fry by using the Universal Steam Oven, select "Fish-High Temperature Grill" mode for 5 minutes.

3. Put green and red bell pepper and celery into the baking pan, sprinkle salt 5g, pour oil 10ml, quick stir-fry by using the Universal Steam Oven, select "Single Point Stratified Grill" mode for 2 minutes.

4. Make the sauce. Pour oil 30ml into a heated wok, add sliced scallion 20g, sliced ginger 20g, sliced garlic 20g, pour soy sauce 15ml, boiled water 1L, then add salt 5g, ground pepper 3g, thicken the soup with water-starch mash 15g, the sauce is done.

5. Mix the vegetable and fish fillets. Pour sauce, stir evenly, sprinkle caraway, the dish is done.

菜品特点

特色 炸烹鱼条与锅塌鱼的制作方法类似，均为炸制，但减少了烧制收汁的过程，料汁较薄，能挂在鱼条上即可。对于鱼肉的选择，以无刺海鱼为佳，海鱼味更鲜美，肉质更细嫩，可以弥补烹饪中缺少收汁过程而导致的味道不足，本菜谱选用的是龙俐鱼，价格较低，非常适合团膳食用。

品味 海鱼味腥，腌制要用葱姜水，可去腥取鲜。料汁咸鲜适口，可以充分突出鱼肉的软嫩鲜香。配菜选用青、红椒和芹菜，口感脆爽，是品味鱼肉的上佳伴侣。

品相 鱼条在腌制时码入淀粉上浆，经过炸制，呈金黄色，咬一口，露出雪白的鱼肉，白金相衬，辅以红绿相间的蔬菜，更凸显了鱼肉的鲜美。

营养价值 龙俐鱼具有海产鱼类在营养上显著的优点，含有较高的不饱和脂肪酸，蛋白质容易消化吸收。其肌肉细嫩，口感爽滑，鱼肉久煮不老，无腥味和异味，属于高蛋白、低脂肪、富含维生素的鱼类。

菜品名称

黄焖龙俐鱼

Name: Braised Long Li Fish

制作人：徐 龙　　中国烹饪大师

Made by: Long Xu　　A Great Master of Chinese Cuisine

主 料 Main Ingredient	调 料 Seasoning	料 酒 Cooking Wine 10ml
龙俐鱼：1000g 切块	清 油 Oil........................60ml	生 抽 Light Soy Sauce......10ml
Long Li Fish 1000g Cut	盐 Salt................................20g	香菜碎 Chopped Caraway....20g
	鸡 蛋 Egg................2 个 (pcs)	葱 片 Sliced Scallion.........20g
配 料 Burdening	胡椒粉 Ground pepper 10g	姜 末 Minced Ginger 20g
胡萝卜：500g 滚刀块	面 粉 Powder....................20g	
Carrot 500g Cut	淀 粉 Starch.....................40g	

制作方法

❶ 首先将龙俐鱼入盆腌制，加盐10g，胡椒粉5g，葱、姜各20g，打入两个鸡蛋，用手搅拌均匀后加面粉20g，淀粉30g，鱼肉均匀裹上面糊后淋上明油50ml，腌制完成。

❷ 烤盘刷底油，将腌制好的鱼块码入烤盘烤熟，选择『单点分层炙烤』模式，时长5分钟。

❸ 烤鱼时调制生汁，碗内放入料酒10ml，盐10g，胡椒粉5g，生抽10ml，淀粉10g，水800ml，搅拌均匀，料汁即成。

❹ 在烤好的鱼块中倒入胡萝卜块，浇上料汁，放入万能蒸烤箱烹制入味，选择『单点分层煎烤—1号色』，时长9分钟，撒上香菜碎，菜品即成。

WORKING PROCESS

1. Marinate the Long Li Fish in a basin. Add salt 10g, ground pepper 5g, scallion 20g, ginger 20g, 2 eggs, stir well and add powder 20g, starch 30g, coated the fish meat with powder and starch mash then pour oil 50ml, the marinating is done.

2. Brush oil at the bottom of the baking pan, baked the fish, select "Single Point Stratified Grill" mode for 5 minutes.

3. Make the sauce at the same time. Pour cooking wine 10ml into a bow, salt 10g, ground pepper 5g, light soy sauce 10ml, starch 10g, water 800ml, stir evenly then the sauce is done.

4. Mix the baked fish with carrot pieces, dress the sauce, steam it by using the Universal Steam Oven, select "Single Point Stratified Grill, color No. 1" for 9 minutes, sprinkle the chopped caraway, the dish is done.

中国大锅菜

菜品名称·黄焖龙俐鱼
Name: Braised Long Li Fish

菜品特点

特色

黄焖鱼是鲁菜经典菜式之一，『焖』是我国传统烹饪技法之一。将锅置于微火之上，将食材慢慢焖至软烂。『焖』在鲁菜中使用较多，根据口味不同，可分为『红焖』与『黄焖』，二者烹调方法和用料相同，只是有些调料用量多寡不一，黄焖在腌制和烹调料汁时放的酱油较少。

品味

鱼肉裹上面糊和明油后，经过烤制，具有滑油的效果，外皮脆爽，肉质鲜嫩。经过焖烧，料汁与鱼块和胡萝卜完美结合，萝卜肉质细密，质地脆嫩，有特殊的甜味，二者将咸鲜之味发挥得淋漓尽致。

品相

黄焖之色区别于红焖，呈浅黄色，料汁几近收干，每一块鱼肉与胡萝卜都挂满芡汁，晶莹诱人，红色的胡萝卜为这道菜更添亮丽，可谓色味俱佳。

营养价值

这道菜咸鲜适口，口味容易调节，制作此菜对鱼的种类不甚挑剔，以刺少的海鱼为佳，龙俐鱼就是上佳之选。龙俐鱼含有较高的不饱和脂肪酸，蛋白质容易消化吸收，其肌肉细嫩，口感爽滑，鱼肉久煮不老，无腥味和异味。胡萝卜营养价值十分丰富，有『小人参』之称，尤其含有丰富的胡萝卜素，维生素C和B族维生素，具有补肝明目，提高人体免疫力的作用。

蓝带炸猪排

Name: LeCordonBleu Style of Deep-fried Pork Tenderloin

制作人：徐 龙　　中国烹饪大师

Made by: Long Xu　　A Great Master of Chinese Cuisine

主 料 Main Ingredient	配 料 Burdening	调 料 Seasoning
猪里脊：1000g　切厚片 Pork Tenderloin　1000g Cut into thick pieces	火 腿：250g　切片 Ham　250g　Sliced 奶 酪：250g　切片 Cheese　250g　Sliced	盐 Salt ....10g 胡椒粉 Ground Pepper ....6g 鸡 蛋 Egg.................4 个 (pcs) 面包糠 Bread Crumbs ....200g 清 油 Oil............................5ml

中国大锅菜

菜品名称·蓝带炸猪排

Name: LeCordonBleu Style of Deep-fried Pork Tenderloin

制作方法

❶ 将猪里脊用肉锤捶打成薄片，加盐10g，胡椒粉6g，腌制少许。

❷ 两片火腿片夹一片奶酪，裹入猪排中，用刀背轻轻压实。

❸ 盆中打入4个鸡蛋，将蛋液打散，然后开始将猪排的『过三关』，将裹入火腿、奶酪的猪排拍粉，一定要拍均匀，使猪排不沾，然后放入蛋液中过一下，再将沾满蛋液的猪排滚上面包糠。如猪排片较厚，需再过蛋液滚上一层面包糠。

❹ 将猪排摆入烤盘中，轻轻刷一层明油，放入万能蒸烤箱烹制成熟，选择『单点分层炙烤~2号色』，时长10分钟，菜品即成。

WORKING PROCESS

1. Beat the pork tenderloin firstly, then add salt 10g, ground pepper 6g, marinate for a while.

2. Stuff a piece of cheese by 2 pieces of ham, then coated them by pork tenderloin. Shape them.

3. Add in 4 eggs into a basin and stir. Then coat the pork tenderloin with powder evenly. Then coat with egg mash. The last step, coat with bread Crumbs.

4. Brush the oil at the surface of the pork in a baking pan. Use the Universal Steam Oven to bake the pork, select "Single Point Stratified Grill, color No. 2" for 10 minutes. The dish is done.

中国大锅菜

蒸烤箱卷（纪念版）

The Big-Wok-Made Cuisine of China, Food Volume of Steam Oven（Commemorative Edition）

菜品特点

特色　这是一道十分地道的法式西餐菜品，蓝带是法国乃至世界最享誉盛名的西餐烹饪学校，是很多人实现名厨梦想的摇篮。蓝带炸猪排是其最经典的菜式之一，制作方法简单，但每一个细节要处理好却并不简单，需要对食材的物性有着良好的把握。

品味　外皮酥香，一口咬下去，肉质十分软嫩，胡椒粉将肉腥味去除，只留下猪肉的香味，奶酪已经融化，香气则更加浓郁，再加上火腿特有的味道，这几样食材混合，都将自己的香味充分展示出来，叫人欲罢不能。

营养价值　猪脊肉含有人体生长发育所需的丰富的优质蛋白、脂肪、维生素等，而且肉质较嫩，易消化。奶酪制品都是由10倍的牛奶浓缩而成，含有丰富的蛋白质、钙、脂肪、磷和维生素等营养成分，能够补充钙质，有利于儿童成长，其热能和脂肪含量均较高，而胆固醇含量低，能够预防心血管疾病。

品相　面包糠经过炸制，外皮呈金黄之色，红嫩的火腿夹着向外流淌着的芝士，诱使人赶紧咬一口，去享受那满口香味。

菜品名称

日式酱烤鱼

Name: Baked Long Li Fish with Japanese Bean Paste

制作人：徐　龙　　中国烹饪大师

Made by: Long Xu　　A Great Master of Chinese Cuisine

主　料　Main Ingredient
龙俐鱼：1000g　切块
Long Li Fish　1000g　Cut

调　料　Seasoning
日本味噌 Japanese Bean

Paste.....................15g

清　酒 Saki 5ml
柠　檬 Lemon ....1 个 (pcs)
紫苏叶 Perilla Leaves
　　　　...............10 片 (pieces)
姜　芽 Ginger Buds..............10g

鸡　蛋 Egg................1 个 (pcs)

中国大锅菜

制作方法

❶ 首先调制酱料，盆中加日本味噌 15g，打入一个鸡蛋黄，倒入清酒 5ml，用手拌匀，然后倒入改刀后的龙俐鱼块，搅拌均匀，腌制 30 分钟。

❷ 烤盘刷底油，将腌制好的龙俐鱼码入盘中，放入万能蒸烤箱烤制成熟，选择『单点分层炙烤』模式，时长 4 分 30 秒。

❸ 盛盘时，将鱼块放在紫苏叶上，码入切好的柠檬块和姜芽，菜品即成。

WORKING PROCESS

1. First of all, make the sauce. Put the Japanese bean paste 15g into a basin, add 1 egg yolk, pour Saki 5ml, stir it evenly. Marinate the Long Li fish with this sauce for 30 minutes.

2. Brush the oil at the bottom of the baking pan, place in the Long Li fish, using the Universal Steam Oven, select "Single Point Stratified Grill" mode for 4 minutesand 30 seconds.

3. Put the fish on Perilla leaves, place the lemon pieces and ginger buds, the dishis done.

中国大锅菜

菜品名称·日式酱烤鱼

Name: Baked Long Li Fish with Japanese Bean Paste

菜品特点

特色 「味噌」翻译成中文就是酱的意思，根据种曲的不同，如米曲、麦曲、豆曲，可以分为不同的类型，也可以按照颜色分类，有白味噌和赤味噌之分，赤味噌发酵时间较长，味道较咸；白味噌发酵时间短，而稍带甜味。这道菜采用的是赤味噌，也叫日本赤酱。

品味 味噌在日本料理中的用途十分广泛，既能够起到良好调味品的作用，又能让食材发挥出其本身的特点。酱料去除了鱼肉的腥味，又在鲜味之外丰富了味道的层次感，是一道品味上佳的美食。提升了其鲜味，

品相 如果是宴会使用，这道菜的摆盘比较讲究，鱼肉置于紫苏叶上，配以芽姜和柠檬，口味上增添一些清淡之感。

营养价值 这道菜宜用刺少的海鱼，本菜单用龙俐鱼。龙俐鱼含有较高的不饱和脂肪酸，蛋白质容易消化吸收，其肌肉细嫩，口感爽滑，鱼肉久煮不老，无腥味和异味。白萝卜含丰富的维生素C和微量元素锌，有助于增强机体的免疫功能，提高抗病能力。

菜品名称

一品素菜

Name: Steamed Radish and Carrot with Chicken Soup

制作人：徐 龙　　中国烹饪大师

Made by: Long Xu　　A Great Master of Chinese Cuisine

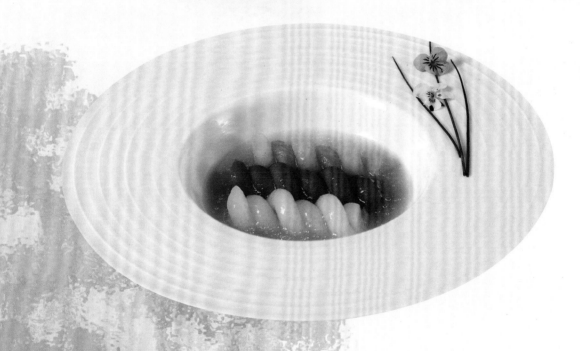

主 料 Main Ingredient

白萝卜：1000g 改刀
White Radish　1000g　Cut into specific shape

青萝卜：1000g 改刀
Green Radish　1000g
Cut into specific shape

胡萝卜：1000g 改刀

Carrot　1000g
Cut into specific shape

调 料　Seasoning

鸡 汤 Chicken Soup 1100ml

盐 Salt 15g

胡椒粉 Ground Pepper 8g

白 糖 Sugar 10g

清 油 Oil 50ml

淀 粉 Starch 20g

香葱段 Chives Sections 10g

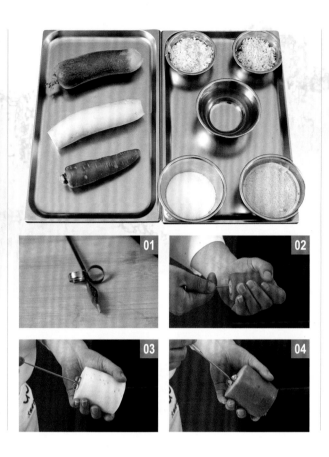

中国大锅菜

菜品名称 · 一品素菜

Name: Steamed Radish and Carrot with Chicken Soup

制作方法

① 将三种萝卜用魔术刀改刀成型，摆入盘中。

② 然后调制生汁，加鸡汤 600ml，盐 10g，胡椒粉 5g，糖 5g，搅拌均匀。菜品放入万能蒸烤箱后浇上生汁，烹制成熟，选择『单点分层蒸煮』模式，时长 15 分钟。

③ 在烹制时炒制芡汁，锅内倒入鸡汤 500ml，锅开后加盐 5g，胡椒粉 3g，白糖 5g，将 20g 淀粉制成的水淀粉拌匀后倒入，不断搅拌，淋上 50ml 明油，芡汁即成。

④ 将芡汁均匀浇在烹制完成的蔬菜上，撒上香葱段，菜品即成。

WORKING PROCESS

1. Cut the 3 kinds of radishes into specific shape.

2. Then make the sauce. Pour chicken soup 600ml, salt 10g, ground pepper 5g, stir it evenly. Put the dish into the Universal Steam Oven and dress the sauce. Select "Single Point Stratified Steam" mode for 15 minutes.

3. To thicken the sauce with starch. Pour chicken soup 500ml, heat till boiled, then add salt 5g, ground pepper 3g, sugar 5g. Turn the starch 20g into water-starch mash, pour it into the wok and stir repeatedly. Pour oil 50ml, the sauce is done.

4. Pour the thickened sauce on the surface of the vegetables, sprinkle chives and the dish is done.

中国大锅菜

蒸烤箱卷（纪念版）

The Big-Wok-Made Cuisine of China, Food Volume of Steam Oven（Commemorative Edition）

菜品特点

特色 素菜也能做得这么有品质，这才是大师的手笔。这道菜食材简单，胜在造型，采用魔术刀，将食材塑造成罗圈形，甚至可以拆开，将两种不同的食材组合在一起，色彩斑斓。

品味 这对于素食主义者是一道美味，采用了三种萝卜，用少许调味品调制出清淡的料汁，既不掩盖萝卜淡淡的香气，又对素味是一很好的调剂。

品相 三种萝卜三种颜色，色彩丰富，叠加在一起具有清新之美，料汁颜色很浅，更给食材添加晶莹之感。

营养价值 萝卜生食、熟食均可，可做汤、可炖、可炒、可盐腌和制作泡菜，也可水煮后切碎做馅。萝卜含有能诱导人体自身产生干扰素的多种微量元素，可增强机体免疫力，并能抑制癌细胞的生长，对防癌、抗癌有重要意义。常吃萝卜可降低血脂、软化血管、稳定血压，预防冠心病、动脉硬化、胆石症等疾病。中医认为，萝卜生食辛甘而性凉，熟食味甘性平，有顺气、宽中、生津、解毒、消积滞、宽胸膈、化痰热、散瘀血之功效。

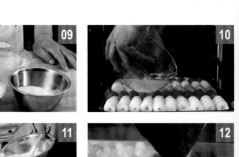

花椒鸡丁

Name: Grilled Chicken with Chinese Prickly Ash

制作人：张爱强　　中国烹饪大师

Made by: Aiqiang Zhang　　A Great Master of Chinese Cuisine

主　料 Main Ingredient
鸡　丁：1000g　切丁
Chicken　1000g　Pieced

调　料 Seasoning
清　油 Oil.........................50ml
花　椒 Chinese Prickly Ash ..20g
盐 Salt.................................20g

胡椒粉 Ground Pepper.........10g
料　酒 Cooking Wine 25ml
白　糖 Sugar.......................10g
酱　油 Soy Sauce 10ml
上　汤 Clear Soup............300ml
干辣椒 Dried Chili 20g
葱　片 Sliced Scallion...........15g
姜　片 Sliced Ginger 15g

备　注 Tips
花椒如果炸煳会有苦味，爆香花椒时火要小一些。
Use light fire during stir-frying the Chinese prickly ash, otherwise the it will taste bitter.

中国大锅菜

The Big-Wok-Made Cuisine of China, Food Volume of Steam Oven（Commemorative Edition）

蒸烤箱卷（纪念版）

制作方法

❶ 首先腌制鸡丁，鸡丁入盆，倒入料酒 15ml，加胡椒粉 5g，盐 10g，搅拌均匀，腌制 15 分钟即可。

❷ 将腌制好的鸡丁放入万能蒸烤箱烤制，选择『单点分层炙烤』模式，时长 5 分钟。

❸ 烤制鸡丁时炒制料汁，锅热加油 50ml，加花椒 20g 爆出香味，主要花椒不要炒煳，滗出一半花椒，加入干辣椒 20g，葱、姜片各 15g 炒香，再加料酒 10ml，酱油 10ml，白糖 10g，倒入上汤 300ml，锅开后加盐 10g，胡椒粉 5g，料汁即成。

❹ 将料汁倒入烤好的鸡丁中，搅拌均匀，菜品即成。

WORKING PROCESS

1. Marinate the chicken in a basin. Pour cooking wine 15ml, ground pepper 5g, salt 10g, well stir and marinate for 15 minutes.

2. Bake the marinated chicken by using the Universal Steam Oven, select "Single Point Stratified Grill" mode for 5 minutes.

3. Make the sauce at the same time. Pour oil 50ml in a heated wok, add Chinese prickly ash 20g, deep-fry till fragrance. Filter half of the Chinese prickly ash and then add dried chili 20g, sliced scallion 15g, sliced ginger 15g, stir-fry and then pour cooking wine 10ml, soy sauce 10ml, sugar 10g, clear soup 300ml. Add salt 10g while the soup boiled, ground pepper 5g, the sauce is done.

4. Pour the sauce into the chicken. The dish is done.

中国大锅菜

菜品名称·花椒鸡丁

Name: Grilled Chicken with Chinese Prickly Ash

菜品特点

特色 这是一道四川家常菜，在明代辣椒传入中国之前，四川等地之人的辛味调味品主要为花椒。花椒味麻，带有一股诱人的香气，有助于健胃祛湿，十分受冬季湿冷地区的人们喜爱。今天，人们在传统基础上又加入辣椒，尤其以使用香气十足的二荆条辣椒为佳。

品味 这是一道典型的川菜，具有麻辣鲜香的特点，鸡肉肉质鲜嫩，经过腌制去除腥味，配上香味浓郁的料汁，麻辣扣敲着味蕾，每一粒鸡丁的入口，都不断打开胃口，是上好的下饭菜。

品相 菜品呈亮红色，辣椒的红色宣告着味道的火辣。这种品相，喜麻辣口味的人是一定不能拒绝的。

营养价值 鸡肉肉质细嫩，滋味鲜美，并富有营养，有滋补养身的作用。鸡肉中蛋白质的含量比例很高，而且消化率高，很容易被人体吸收利用，有增强体力、强壮身体的作用。中医认为鸡肉性平、温、味甘，入脾、胃经，可益气、补精、添髓。

菜品名称

小 煎 鸡

Name: Small Fried Chicken

制作人：张爱强　　中国烹饪大师

Made by: Aiqiang Zhang　　A Great Master of Chinese Cuisine

主 料 Main Ingredient
鸡腿肉：2000g　切丁
Drumstick　2000g　Cut

配 料 Burdening
青 笋：300g　切条
Green Bamboo Shoot　300g
Striped

芹 菜：250g　切条
Celery　250g　Striped

调 料 Seasoning
盐 Salt 10g
清 油 Oil 80ml
淀 粉 Starch......................20g
胡椒粉 Ground Pepper 3g

酱 油 Soy sauce 10ml
泡 椒 Soaked Chili............100g
料 酒 Cooking Wine 20ml
糖 Sugar 5g
葱 片 Sliced Scallion..........15g
姜 片 Sliced Ginger 15g

中国大锅菜

菜品名称·小煎鸡
Name: Small Fried Chicken

制作方法

① 首先腌制鸡肉，加盐 10g，料酒 10ml，淀粉 10g，胡椒粉 3g，酱油 5ml，搅拌均匀，腌制 20 分钟。

② 烤盘刷底油，将腌制好的鸡块放入万能蒸烤箱滑油，选择『单点分层炙烤』模式，时长 5 分钟。

③ 滑油时烧制料汁，锅热下油 80ml，加入泡椒 100g，爆香后加青笋 200g，葱、姜片各 15g，倒入清水 200ml，锅开后加糖 5g，料酒 10ml，酱油 5ml，淀粉 10g 制成的水淀粉，芹菜条 100g，勾好芡后料汁即成。

④ 将料汁倒入滑油后的鸡块中，趁热搅拌均匀，菜品即成。

WORKING PROCESS

1. Marinate the drumstick at first. Add salt 10g, cooking wine 10ml, starch 10g, ground pepper 3g, soy sauce 5ml, stir well and marinate for 20 minutes.

2. Brush oil at the bottom of the baking pan, place drumstick at the baking pan and quick bake by using the Universal Steam Oven, select "Single Point Stratified Grill" mode for 5 minutes.

3. Make the sauce at the same time. Pour oil 80ml into a heated wok, then add soaked chili 100g, deep-fry and add green bamboo shoot 200g, sliced scallion 15g, sliced ginger 15g, pour water 200ml. Add sugar 5g while the water is boiled, cooking wine 10ml, soy sauce 5ml, water-starch mash 10g, striped celery 100g, thicken the soup and the sauce is done.

4. Well stir the sauce and the drumstick, the dish is done.

中国大锅菜

蒸烤箱卷（纪念版）

The Big-Wok-Made Cuisine of China, Food Volume of Steam Oven (Commemorative Edition)

菜品特点

特色 小煎鸡是一道川菜家常菜，相传起源于自贡，自贡古为盐都，盛产井盐，十分富庶，酒肆林立，赫赫有名的『盐帮菜』诞生于此，以口味厚重著称。这道菜以采用童子鸡为佳，肉质更细嫩，更加鲜美，喜欢吃辣的人士可以在烹调中加入二荆条辣椒，菜品会更加鲜香。

品味 泡菜是川菜的灵魂，控制这道菜口味的就是泡椒，泡椒具有色泽红亮，辣而不燥、辣中微酸的特点，加上鸡肉自有的鲜香、细嫩，和酱油、葱、姜一起构成了复合口味，美味之中带有一股温文尔雅的辣味。

品相 这道菜色泽红亮，红绿相映，鸡肉上挂满薄薄的芡汁，菜色诱人，泡椒的香气更是让人味蕾大开。

营养价值 鸡肉肉质细嫩，滋味鲜美，并富有营养，有滋补养身的作用。鸡肉中蛋白质的含量比例很高，而且消化率高，很容易被人体吸收利用，有增强体力、强壮身体的作用。青笋的营养成分很多，包括蛋白质、脂肪、糖类、灰分、维生素A原、维生素B₁、维生素B₂、维生素C、钙、磷、铁、钾、镁、硅等和食物纤维，故可增进骨骼、毛发、皮肤的发育，有助于人的生长。旱芹含有丰富的维生素A、维生素B₁、维生素B₂、维生素C和维生素P、钙、铁、磷等矿物质含量也多，此外还有蛋白质、甘露醇和食物纤维等成分，具有安神、利尿、消肿之功效。

菜品名称

盐 煎 肉

Name: Salty-fried Pork

制作人：张爱强　　中国烹饪大师

Made by: Aiqiang Zhang　　A Great Master of Chinese Cuisine

主 料 Main Ingredient
五花肉：1500g　切薄片
Pork Belly　1500g　Sliced

配 料 Burdening
青 蒜：100g　切段
Garlic Sprouts　100g　Cut

调 料 Seasoning
清 油 Oil.......................100ml
泡椒碎 Chopped Soaked
　　　　Chili 30g
郫县豆瓣酱 Pixian Chili Bean
　　　　Paste.................30g
豆 豉 Fermented Soya
　　　　Beans 30g

料 酒 Cooking Wine 10ml
甜面酱 Sweet Bean Sauce.....20g
酱 油 Soy Sauce 10ml
醋 Vinegar............................5ml
盐 Salt 10g
料 酒 Cooking Wine 20ml

中国大锅菜

蒸烤箱卷（纪念版）

The Big-Wok-Made Cuisine of China, Food Volume of Steam Oven（Commemorative Edition）

制作方法

❶ 将五花肉片加盐10g，料酒20ml，腌制15分钟。

❷ 烤盘刷油，将五花肉切薄片，码盘干煸，选择『单点分层炙烤』模式，时长9分钟。

❸ 炒制料汁，锅热下油100ml，加入泡椒碎30g，郫县豆瓣酱30g，豆豉30g，炒香，烹入料酒10ml，甜面酱20g，酱油10ml，醋5ml，倒入青蒜，料汁即成。

❹ 将料汁趁热拌入刚出锅的肉片中，搅拌均匀，菜品即成。

WORKING PROCESS

1. Marinate the sliced pork belly with salt 10g, cooking wine 20ml for15 minutes.

2. Brush oil on baking pan, slice the pork belly and deep-fry, select "Single Point Stratified Grill" mode for 9 minutes.

3. Make the sauce. Pour oil 100ml into a heated wok, add chopped soaked chili 30g, Pixian chili bean paste 30g, fermented soya beans 30g, stir-fry till fragrance. Pour cooking wine 10ml, sweet bean sauce 20g, soy sauce 10ml, vinegar 5ml, sprinkle garlic sprouts. The sauceis done.

4. Mix the meat with the sauce. Well stir and the dish is done.

菜 品 特 点

特色 这是一道四川的家常菜，口味上与回锅肉相似，并称姐妹菜，是下酒佐餐的佳肴。

品味 菜品具有川菜麻辣鲜香的特点，五花肉干煎而熟，略焦，微脆，带有肉的干香，辅以料汁，口味厚重，油而不腻，辣而不燥，回味无穷。

品相 菜品色泽红亮，呈深红色，肉片上挂满料汁，油亮诱人。

营养价值 五花肉位于猪的腹部，猪腹部脂肪组织很多，其中又夹带着肌肉组织，肥瘦间隔，故称『五花肉』。其含有丰富的优质蛋白质和必需的脂肪酸，并提供血红素（有机铁）和促进铁吸收的半胱氨酸，能改善缺铁性贫血；具有补肾养血，滋阴润燥的功效。

鱼香八块鸡

Name: Baked Drumsticks with Fish flavor Sauce

制作人：张爱强　　中国烹饪大师

Made by: Aiqiang Zhang　　A Great Master of Chinese Cuisine

主　料　Main Ingredient
鸡腿肉：2000g　切大块
Drumstick　2000g　Cut

调　料　Seasoning
清　油　Oil........................100ml
郫县豆瓣酱　Pixian Chili Bean
　　　　Paste.................50g

泡椒碎　Chopped Soaked
　　　Chili..........................80g
料　酒　Cooking Wine.........30ml
盐　Salt.................................10g
胡椒粉　Ground Pepper..........5g
酱　油　Soy Sauce................5ml
醋　Vinegar...........................30ml
糖　Sugar..............................30g

淀　粉　Starch.......................20g
葱　Scallion............................50g
姜　Ginger..............................50g

中国大锅菜

菜品名称·鱼香八块鸡

Name: Baked Drumsticks with Fish flavor Sauce

制作方法

① 首先将鸡腿肉块入盆腌制，加葱、姜片各30g，盐10g，胡椒粉5g，倒入料酒20m¹，搅拌均匀，腌制15分钟。

② 将腌制好的鸡块放入万能蒸烤箱烤熟，选择『单点分层炙烤』模式，时长9分钟。

③ 烧制鱼香汁，锅热下油100m¹，油热下郫县豆瓣酱50g，泡椒碎80g，葱、姜末各20g爆香，烧出红油后，烹入料酒10m¹，酱油5m¹，醋30m¹，白糖30g，倒入开水200m¹，锅开后倒入水淀粉20g勾芡。

④ 将料汁浇在鸡块上，拌匀，放入万能蒸烤箱再次烹制入味，选择『单点分层煎烤』模式，时长3分钟，菜品即成。

WORKING PROCESS

1. Marinate the drumsticks in a basin. Add sliced scallion 30g, sliced ginger 30g, salt 10g, ground pepper 5g, pour cooking wine 20ml, well stir and marinate for 15 minutes.

2. Bake the drumsticks by using the Universal Steam Oven, select "Single Point Stratified Grill" mode for 9 minutes.

3. Make the fish flavor sauce. Pour oil 100ml into a heated wok, then add Pixian chili bean paste 50g, chopped soaked chili 80g, minced scallion 20g, minced ginger 20g, deep-fry. Then pour cooking wine 10ml, soy sauce 5ml, vinegar 30ml, sugar 30g, boiled water 200ml, then thicken the soup by mixing the water-starch mash 20g.

4. Dressing the sauce on the chicken, stir evenly. Bake by using the Universal Steam Oven, select "Single Point Stratified Bake" mode for 3 minutes, the dish is done.

中国大锅菜

蒸烤箱卷（纪念版）

The Big-Wok-Made Cuisine of China, Food Volume of Steam Oven（Commemorative Edition）

菜品特点

特色 鱼香八块鸡属川菜，由鱼香肉丝演变而来，为鱼香味型。鱼香味型的核心就是泡椒，泡椒又称『鱼辣子』，在川菜烹调中广为应用，是烹制鱼香味菜肴必不可少的调味料。

品味 这道菜中虽无鱼，却能吃出鱼的鲜味，并且咸甜酸辣兼具。葱姜香气浓郁，是典型的复合香味。鸡肉经过烤制，外焦里嫩，外皮的干脆更易吸收料汁，十分入味，料汁的鲜香和质嫩的鸡肉相得益彰，是下饭佳肴。

品相 这道菜色泽红亮，鸡肉上挂满料汁，葱姜末附着其上，看上去十分入味，鱼香味远远可闻，色香诱人。

营养价值 鸡肉肉质细嫩，滋味鲜美，并富有营养，有滋补养身的作用。鸡肉中蛋白质的含量比例很高，而且消化率高，很容易被人体吸收利用，有增强体力、强壮身体的作用。中医认为鸡肉性平、温、味甘、入脾、胃经，可益气、补精、添髓。

菜品名称

辣子鸡丁

Name: Stir-fried Diced Chicken with Chili

制作人：张伟利　　中国烹饪大师

Made by: Weili Zhang　　A Great Master of Chinese Cuisine

主 料 Main Ingredient	鲜红辣椒：250g　切丁
鸡腿肉：2500g　切丁	Fresh Red Chili　250g　Diced
Drumstick　2500g　Diced	

配 料 Burdening
黄 瓜：500g　切丁
Cucumber　500g　Diced
胡萝卜：500g　切丁
Carrot　500g　Diced

调 料 Seasoning
清 油 Oil 300ml
盐 Salt 40g
白 糖 Sugar 20g
味 精 Monosodium
　　　 Glutamate 5g

淀 粉 Starch 100g
酱 油 Soy Sauce 30ml
料 酒 Cooking Wine 20ml
郫县豆瓣酱 Pixian Chili Bean
　　　　　 Paste 100g
鸡 蛋 Egg .... 2 个 (pcs)
姜 末 Minced Ginger 10g
葱 末 Minced Scallion 10g
蒜 末 Minced Garlic 10g

中国大锅菜

The Big-Wok-Made Cuisine of China, Food Volume of Steam Oven（Commemorative Edition）

蒸烤箱卷（纪念版）

制作方法

❶ 首先将鸡肉放入万能蒸烤箱飞水，选择『蒸制蔬菜』模式，99℃，时长3分钟。同时将黄瓜丁和胡萝卜丁焯水，选择『蒸制蔬菜』模式，99℃，黄瓜丁1分钟，胡萝卜丁3分钟。将100g淀粉制成水淀粉。

❷ 将飞水后的鸡丁放入盆中腌制，加盐10g，打入2个鸡蛋，加一半水淀粉，腌制15分钟。

❸ 烤盘内刷底油，将腌制后的鸡丁放入烤盘，用万能蒸烤箱滑油，选择『高温炙烤』模式，时长3分钟。

❹ 在滑油时淋汁，锅热下油300ml，油热加郫县豆瓣酱100g，炒出红油，放入葱、姜、蒜末各10g炒香，再加白糖20g，盐30g，酱油30ml，料酒20ml，味精5g，加水750ml烧开，加入剩余水淀粉勾芡，料汁既成。

❺ 将加工好的主辅料盛入布菲芯盆中，浇上料汁拌匀，撒上红辣椒，即可出锅。

WORKING PROCESS

1. Steam the drumstick by using the Universal Steam Oven, select "Steam Vegetable" mode, 99℃ ,for 3 minutes. At the same time quick-boil the diced cucumber and diced carrot, select "Steam Vegetable" mode, 99℃ , 1 minute for cucumber and 3 minutes for carrot. Turn the starch 100g into water-starch mash.

2. Marinate the drumstick in a basin after steaming, add salt 10g, plus 2 eggs, water-starch mash (the 1st half), marinate for 15 minutes.

3. Brush oil at the bottom of the baking pan, put the drumstick into baking pan, grill by using the Universal Steam Oven, select "High Temperature Grill" mode for 3 minutes.

4. Make the sauce at the same time, pour oil 300ml into a heated wok, then add Pixian chili bean paste 100g, stir-fry till red oil comes out, then add minced scallion 10g, minced ginger 10g, minced garlic 10g, stir-fry till fragrance, then add sugar 20g, salt 30g, soy sauce 30ml, cooking wine 20ml, MSG 5g, water 750ml, heat the water till boiled, then add water-starch mash (the 2nd half) to thicken the sauce, then the sauce will be ready.

5. Put all ingredients into a buffet pot, pour the sauce, the dish is done.

中国大锅菜

菜品名称・辣子鸡丁

Name: Stir-fried Diced Chicken with Chili

菜品特点

特色 辣子鸡丁属于川菜系，起源于川东，是一道著名的江湖菜。这道菜流传很广，倍受广大食客的喜爱，在流传过程中也有所改变，进一步适应当地的饮食习惯。团体膳食制作此菜，要对辣度有所调整，以适应大部分人的口味。

品味 本菜谱所收的这道辣子鸡丁是用郫县豆瓣酱炒制的，相比之下，辣度更低，而突出了其鲜香。这道菜口味较重，但又不是一味地辣，而是香气十足，配菜选用黄瓜与胡萝卜，清脆爽口，有助于缓解辣椒所带来的灼烧感。

品相 红红的辣椒与金灿灿的鸡丁是这道菜的主旋律，整道菜色泽红亮，中间点缀黄瓜与胡萝卜丁，火辣之余又给人耳目一新的感觉。

营养价值 川菜以重油重麻辣著称，这道菜能够极大地调动食客的胃口。鸡肉性平、温、味甘，入脾、胃经，可益气、补精、添髓。而鸡腿肉的胶原蛋白含量十分丰富，对于保持皮肤光泽、增强皮肤弹性均有好处。黄瓜富含维生素，中医认为其性凉，辣椒的火辣对其有所中和。

菜品名称

香辣鳕鱼

Name: Grilled Spicy-Flavor Codfish

制作人：张伟利　　中国烹饪大师

Made by: Weili Zhang　　A Great Master of Chinese Cuisine

主　料　Main Ingredient
鳕　鱼：4000g　切块
Codfish　4000g　Cut

配　料　Burdening
洋　葱：250g　切丝
Onion　250g　Shredded
香　菜：100g　切段
Caraway　100g　Cut

调　料　Seasoning
清　油　Oil...........................30ml
盐　Salt.................................20g
料　酒　Cooking wine..........15ml
料　油　Spicing Oil 10ml
孜　然　Cumin........................5g
蚝　油　Oyster Sauce............30g
酱　油　Soy Sauce..............30ml
白　糖　Sugar.......................20g
郫县豆瓣酱　Pixian Chili Bean
　　　　　　Paste.................50g

辣椒面　Ground Chili..............10g
五香粉　Five-Flavor Powder 5g
蒜　　末　Minced Garlic............30g
姜　　末　Minced Ginger 20g

备　注　Tips
腌制手法要抓揉，将蔬菜中的汁液抓出，使鳕鱼更易吸收。
Knead the codfish during marinating which helps the codfish to absorb the sauce.

中国大锅菜

菜品名称 · 香辣鳕鱼
Name: Grilled Spicy-Flavor Codfish

制作方法

❶ 首先腌制鳕鱼，将鳕鱼放入盆中，加洋葱丝250g，香菜段100g，糖20g，盐20g，郫县豆瓣酱50g，蒜末30g，姜末20g，孜然5g，辣椒面10g，五香粉5g，酱油30ml，料酒15ml，蚝油30g，用抓揉的手法拌匀，腌制1小时。

❷ 烤盘刷底油，将料油10ml拌入腌制好的鳕鱼，摆入烤盘，选择『高温炙烤』模式，时长7分钟。

WORKING PROCESS

1. Marinate the codfish for 1 hour. Put the codfish into a basin, add shredded onion 250g, caraway 100g, sugar 20g, salt 20g, Pixian chili bean paste 50g, minced garlic 30g, minced ginger 20g, cumin 5g, ground chili 10g, five-flavor 5g, soy sauce 30ml, cooking wine 15ml, oyster sauce 30g, stir evenly.

2. Brush the oil at the bottom of the baking pan, push the cooking wine 10ml on marinated codfish, select "High Temperature Grill" mode for 7 minutes.

中国大锅菜

蒸烤箱卷（纪念版）

The Big-Wok-Made Cuisine of China, Food Volume of Steam Oven（Commemorative Edition）

菜品特点

特色 这道菜为烤制菜品，结合川菜调味料，溶入香辣元素，使得烤鳕鱼这道菜品别具一番风味。鳕鱼是西方各国人民最喜爱的鱼类之一，以前我国很少食用，近些年方才十分流行。

品味 鳕鱼需要腌制较长时间，辣味深入鱼肉之中，经过烤制后，香辣可口。鳕鱼肉质雪白，紧致成型，少有腥膻之味，十分适合烤制食用，外皮略焦，而内里软嫩，香辣可口，又不掩鱼肉的鲜美。

品相 腌制后烤熟的鳕鱼块带有酱红色，上面附着少许辣椒，十分入味，给人以十足的食欲。咬一口，鱼肉的白嫩便露了出来，展示着鱼肉的鲜美。

营养价值 鳕鱼蛋白质含量非常高，而脂肪含量极低，至少低于 0.5%。更重要的是，鳕鱼的肝脏含油量高达 45%，同时含有 A、D 和 E 等多种维生素。鱼肉中含有丰富的镁元素，对心血管系统有很好的保护作用，有利于预防高血压、心肌梗死等心血管疾病。鳕鱼低脂肪、高蛋白，低胆固醇、高营养，易于被人体吸收，是老少皆宜的营养食品。

菜品名称

鱼香肉丝

Name: Braised Pork Tenderloin

制作人：张伟利　　中国烹饪大师

Made by: Weili Zhang　　A Great Master of Chinese Cuisine

主　料 Main Ingredient
猪里脊: 1000g　切丝
Pork Tenderloin　1000g　Shredded

配　料 Burdening
青椒: 500g　切丝
Green Bell Pepper　500g
Shredded
红椒: 250g　切丝
Red Bell Pepper　250g　Shredded

木　耳: 250g　切丝
Black Fungus　250g　Shredded

调　料 Seasoning
清　油 Oil.........................200ml
盐 Salt...................................30g
料　酒 Cooking Wine.........20ml
味　精 Monosodium
　　　　Glutamate.................10g
酱　油 Soy Sauce..............30ml

醋 Vinegar..........................30ml
白　糖 Sugar.......................30g
淀　粉 Starch.......................50g
鸡　蛋 Egg................2个（pcs)
郫县豆瓣酱 Pixian Chili Bean
　　　　　　Paste...............200g
姜　末 Minced Ginger..........40g
蒜　末 Minced Garlic............40g
葱　末 Minced Scallion.........40g

中国大锅菜

蒸烤箱卷（纪念版）

The Big-Wok-Made Cuisine of China, Food Volume of Steam Oven（Commemorative Edition）

制作方法

❶ 将青红椒丝和木耳丝放入万能蒸烤箱焯水，选择『单点分层炙烤』模式两分钟。

❷ 将肉丝放入盆中进行腌制，加盐15g，鸡蛋两个，淀粉20g制成的水淀粉，拌匀后放入万能蒸烤箱滑油。

❸ 在滑油时候进行炒汁，锅热下油200ml，油热放入郫县豆瓣酱200g，炒出红油，下葱、姜、蒜末各30g，煸炒出香味，再加酱油30ml，料酒20ml，醋30ml，白糖30g，盐15g，味精10g，水300ml，烧开后加水淀粉30g勾芡，出锅前再放入葱、姜、蒜末各10g，料汁即成。

❹ 将滑油后的肉丝和焯水后的青红椒丝、木耳丝盛入布菲芯盒中，倒入料汁，搅拌均匀，即可。

WORKING PROCESS

1. Steam the shredded meat via the Universal Steam Oven, select "Single Point Stratified Steam" mode for 3minutes. Blanch the shredded green and red bell pepper and shredded black fungus by using the Universal Steam Oven, select "Single Point Stratified Grill" mode for 2 minutes.

2. Marinating the meat after steam, add salt 15g, 2 eggs. Turn the starch 20g into water-starch mash, well stir and bake by the Universal Steam Oven. Select "Single Point Stratified Grill" mode for 2 minutes.

3. Make the sauce at the same time, pour oil 200ml into a heated wok, add Pixian chili bean paste 200g, stir-fry till coming out the red oil, add minced scallion 30g, minced ginger 30g, minced garlic 30g, stir-fry till fragrance, then add soy sauce 30ml, cooking wine 20ml, vinegar 30ml, sugar 30g, salt 15g, MSG 10g, water 300ml, add water-starch mash after the water boiled to thicken the sauce, then add minced scallion 10g, minced ginger 10g, minced garlic 10g, the sauce is done.

4. Put all the ingredients into buffet pot, pour the sauce, well stir, the dish is done.

中国大锅菜

菜品名称·鱼香肉丝
Name: Braised Pork Tenderloin

菜 品 特 点

特色

这是一道传统川菜名菜，流传很广，受到百姓喜爱，在『神舟』载人航天飞船带上太空的航天员食品中，就有鱼香肉丝。据专家考证，这道菜形成于20世纪初，关于这道菜还有一个典故：一个家庭主妇将烧鱼剩下的调料用来炒菜，无意间却发现了一种异乎寻常的口味，十分美味，就是今天鱼香肉丝的雏形。鱼香肉丝的形成深深根植于四川的传统烹饪美食，其中很多原材料皆以四川本地所产为佳，比如泡菜，蜀人善于腌制泡菜，用以烹饪或者直接食用，以泡菜为基础，形成了很多菜肴，鱼香肉丝就是其中之一。

品味

鱼香肉丝的口味是没有鱼肉，而能够品尝出鱼肉的味道，其实形容它的味道，用东晋时期一本描写蜀中文化之书华阳国志中的一句话最恰当：『尚滋味、好辛香』。鱼香肉丝是典型的复合香味，炒制它要使用多种调味料，糖和醋的搭配使其滋味鲜香无比，而辣椒则产生辛香之味，使其味道又增色几分。

品相

鱼香肉丝以猪里脊丝为主料，辅料的使用可以根据实际情况进行调整，不同辅料产生的品相上的差异还是很大的。本菜谱选用青椒、红椒和木耳做辅料，金黄的肉丝上裹着亮红的红油，青椒和木耳则使其色彩和口感更加丰富，观其颜色，闻其味道，令人垂涎欲滴。

营养价值

这是一道开胃又下饭的菜肴，又重油重味，是补充能量的佳选。猪肉是目前人们餐桌上重要的动物性食品之一，猪脊肉含有人体生长的发育所需的丰富的优质蛋白、脂肪、维生素等，而且肉质较嫩，易消化。辣椒含有丰富的维生素C，含有较多抗氧化物质，能改善食欲、增加饭量，辣椒还具有强烈的促进血液循环的作用，可以改善怕冷、冻伤、血管性头痛等症状。木耳含有大量的碳水化合物、蛋白质、铁、钙、磷、胡萝卜素、维生素等营养物质，具有益气、润肺、补脑、轻身、凉血、止血、涩肠、活血、养颜等功效。

菜品名称

孜然鸭腿

Name: Baked Duck-Leg with Cumin

制作人：张伟利　　中国烹饪大师

Made by: Weili Zhang　　A Great Master of Chinese Cuisine

主　料　Main Ingredient
鸭　腿：4000g　整腿
Duck-leg　4000g　The original

配　料　Burdening
洋　葱：500g　切丝
Onion　500g　Shredded
香　菜：150g　切段
Caraway　150g　Cut

调　料　Seasoning
清　油　Oil..........................30ml
盐　Salt 20g
料　酒　Cooking Wine 30ml
孜　然　Cumin......................10g
辣椒面　Ground Chili...............5g
老　抽　Dark Soy Sauce........5ml
姜　末　Minced Ginger 20g
葱　末　Minced Scallion.........20g

蒜　末　Minced Garlic............20g

备　注　Tips
腌制手法要抓揉，将蔬菜中的汁液抓出，使鸭腿更易吸收。
Knead the duck-leg to make the meat oil absorbs the vegetable sauce easily.

中国大锅菜

菜品名称·孜然鸭腿
Name: Baked Duck-Leg with Cumin

制作方法

❶ 首先腌制鸭腿，将鸭腿放入盆中，加盐20g，葱、姜、蒜末各20g，洋葱丝500g，香菜段150g，料酒30ml，糖20g，辣椒面5g，孜然10g，老抽5ml，然后抓揉均匀，腌制1小时。

❷ 烤盘刷底油，将腌制好的鸭腿摆入烤盘，选择「煎烤」模式，时长15分钟。

WORKING PROCESS

1. Marinate the duck leg. Put into a basin, add salt 20g, minced scallion 20g, minced ginger 20g, minced garlic 20g, shredded onion 500g, caraway 150g, cooking wine 30ml, sugar 20g, ground chili 5g, cumin 10g, dark soy sauce 5ml, knead and stir it evenly, marinating for 1 hour.

2. Brush the oil at the bottom of a baking pan, put the marinated duck leg into the baking pan, select "Grill" mode for 15 minutes.

中国大锅菜

蒸烤箱卷（纪念版）

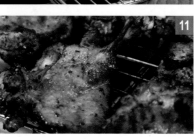

菜品特点

特色 这是一道烤制菜品，鸭子是我国百姓十分喜爱的肉食，尤其是南方地区，更加流行。

品味 这道菜味道的灵魂就是孜然，孜然附着在鸭腿上，经过烤制，产生一种十分诱人的香味，就上一点辣椒面，口感微辣，打开味蕾。鸭肉经过烤制，外皮酥脆，内里软嫩，经过较长时间的腌制，亦十分入味。

品相 鸭肉外皮焦黄，外焦而里嫩，孜然和辣椒烤得干香诱人。由于加入了糖和老抽，色泽会微微有点深，看上去更加入味。

营养价值 鸭肉性凉，经过腌制和烤制，可以在很大程度上中和鸭肉的凉性。鸭肉中含有丰富的蛋白质，容易被人体吸收，所含B族维生素和维生素E较其他肉类多，能有效抵抗脚气病，神经炎和多种炎症，还能抗衰老。

菜品名称

宫保豆腐

Name: Kung Pao Tofu

制作人：赵春源　中国烹饪大师

Made by: Chunyuan Zhao　A Great Master of Chinese Cuisine

主 料 Main Ingredient
豆 腐：2000g 切丁
Tofu 2000g Pieced

配 料 Burdening
胡萝卜：1000g 切丁
Carrot 1000g Pieced
黄 瓜：500g 切丁
Cucumber 500g Pieced
花 生：500g 去皮

Peanut 500g Peeled

小 料 Other Seasoning
葱：50g 切节
Scallion 50g Cut

调 料 Seasoning
清 油 Oil........................100ml
盐 Salt 20g
酱 油 Soy Sauce 10ml

白 糖 Sugar........................80g
料 酒 Cooking Wine 50ml
花 椒 Pepper.......................5g
干辣椒 Dried Chili 30g
郫县豆瓣酱 Pixian Chili Bean
　　　　　Paste.................50g
醋 Vinegar........................150ml
料 油 Spicing Oil 30ml
淀 粉 Starch.......................50g

中国大锅菜

The Big-Wok-Made Cuisine of China, Food Volume of Steam Oven（Commemorative Edition）

蒸烤箱卷（纪念版）

制作方法

❶ 将豆腐炸制，烤盘内刷底油，将豆腐码入烤盘，上面刷一层明油，放入万能蒸烤箱，选择『单点分层炙烤』模式，时长10分钟，待豆腐变成金黄色即可。

❷ 将胡萝卜和黄瓜放入万能蒸烤箱飞水，选择『蒸制蔬菜』模式，胡萝卜时长5分钟，黄瓜时长1分钟。将50g淀粉制成水淀粉备用。

❸ 炒制宫保汁，锅热下油100ml，加干辣椒30g，炸香后捞出备用，加花椒5g，料酒50ml，郫县豆瓣酱50g，酱油10ml，盐20g，糖80g，醋150ml，倒入水1L，搅拌均匀，锅开后倒入水淀粉勾芡，加葱节50g调味，淋上料油30ml，料汁即成。

❹ 将炸制好的豆腐和飞水后的胡萝卜、黄瓜倒入布菲盒中，均匀浇上料汁，搅拌均匀，撒上炸制好的辣椒段、花生，即可盛盘。

WORKING PROCESS

1. Deep-fried the tofu firstly, brush oil at the bottom of the baking pan, the tofu should be coded into the baking pan, brush oil on the surface of the Tofu, put it into the Universal Steam Oven, select "Single Point Stratified Grill" mode for 10 minutes till the tofu's color turns into golden yellow.

2. Steam the carrot and cucumber in the Universal Steam Oven, select "Steam Vegetable" mode, steam 5 minutes for carrot and 1 minute for cucumber. Mix the water and starch to make the watch-starch mash, set aside for later use.

3. Make Kung Pao dressing. Pour oil 100ml into a heated wok, deep-fried dried chili 30g till fragrance, add pepper 5g, cooking wine 50ml, Pixian chili bean paste 50g, soy sauce 10ml, salt 20g, sugar 80g, vinegar 150ml, pour water 1L, well stir, add water-starch mash to thicken the dressing after the boiled, sauce it with scallion 50g, pour spicing oil 30ml, then the dressing is done.

4. Put the deep-fried tofu and steamed carrot, cucumber into the buffet pot. Dressing it and well stirred. Then sprinkle the deep-fried chili and peanut. The dish is done.

中国大锅菜

菜品名称·宫保豆腐
Name: Kung Pao Tofu

菜品特点

特色 宫保豆腐是由一道负有盛名的川菜宫保鸡丁演化而来，由豆腐、干辣椒和各种辅料等炒制而成。由于食材简单易得，又适合下饭，因此成为一道广受民间老百姓欢迎的家常菜。

品味 宫保豆腐味道鲜辣香甜，使用万能蒸烤箱炙烤后，外表金黄，内里软嫩，又因增添了辣椒的辣味变得尤为爽口。豆腐和花生米、青椒一软一硬的搭配，互相衬托、互相增益，豆腐的软滑口感因而更加分明。

品相 宫保豆腐色泽红润，块状的豆腐外形饱满，闪耀着酱汁的光泽，辣椒、花生米和胡萝卜、黄瓜的色彩都鲜艳夺目，使这道菜在视觉上就给人以不可抗拒的美感和魅力。

营养价值 豆腐是一种以黄豆为主要原料的食物，不但含有高蛋白，低脂肪，还含有铁、钙、磷、镁和其他人体必需的多种微量元素，素有「植物肉」之美称，具有降血压、降血脂、降胆固醇的功效。辅以胡萝卜、黄瓜等蔬菜，更增添了营养价值。

菜品名称

麻辣鸡条

Name: Deep-fried Spicy Chicken Strips

制作人：赵春源　　中国烹饪大师

Made by: Chunyuan Zhao　　A Great Master of Chinese Cuisine

主　料 Main Ingredient
鸡腿肉：1500g　切条
Drumstick　1500g　Slit

配　料 Burdening
青、红椒：200g　切条
Green/red Bell Pepper　200g　Slit
洋　葱：200g　切条
Onion　200g　Slit
花生碎：50g
Peanut Minced　50g

小　料 Other Seasoning
姜：10g　切末、切丝

Ginger　10g　Minced and Slit
葱：15g　切末、切丝
Scallion　15g　Minced and Slit

调　料 Seasoning
清　油　Oil.........................20ml
盐　Salt.................................20g
孜　然　Cumin........................5g
白　糖　Sugar.......................10g
料　酒　Cooking Wine.........10ml
胡椒粉　Ground Pepper...........5g
麻　椒　Hemp Pepper...........10g
干辣椒　Dried Chili................10g
淀　粉　Starch.......................50g

料　油　Spicing Oil 10ml

备　注 Tips
1. 鸡肉出油较多，盘内底油要少。
2. 麻辣鸡条的汁是干汁，不要加水和水淀粉。

1. Brush less oil at the bottom of baking pan will be fine cause chicken remains fat.
2. Don't add water and water-starch mash into the dressing.

中国大锅菜

制作方法

1. 将葱丝、姜丝泡入水中，制成葱姜水备用。

2. 首先将鸡条入盆腌制，倒入葱姜水 5ml，胡椒粉 5g，料酒 10ml，盐 10g，抓揉搅拌均匀，给鸡肉上劲，然后打入 1 个鸡蛋，放入淀粉 50g，抓匀后倒入料油 10ml，腌制 15 分钟。

3. 将腌制好的鸡条放入万能蒸烤箱滑油，烤盘内刷少许底油，鸡片码入烤盘中，选择『单点分层炙烤』模式，时长 9 分钟。

4. 在滑油同时，炒制料汁。锅热下油 20ml，加入麻椒 10g，煸炒出香味。加葱末、姜末各 10g，干辣椒段 10g，炒香后加盐 10g，糖 10g，搅拌均匀，加青、红椒条和洋葱，煸炒一会，加花生碎 50g，孜然 5g，搅拌均匀，料汁即成。

5. 将料汁拌入滑好油的鸡条，即可盛盘。

WORKING PROCESS

1. Soak the minced scallion and ginger in water and set aside for later use.

2. Sauce the chicken strips in a pot, pour scallion-ginger water 5ml, ground pepper 5g, cooking wine 10ml, salt 10g. Knead it and stir well. Then add 1 egg, starch 50g, grasp evenly and pour spicing oil 10ml, sauce it for 15 minutes.

3. Slightly bake the sauced chicken strips in the Universal Steam Oven, brush a few oil at the bottom of the baking pan, select "Single Point Stratified Bake" mode for 9 minutes.

4. Make the dressing at the same time. Pour oil 20ml in a heated wok, add hemp peeper 10g and stir-fried till fragrance. Add minced scallion 10g, minced ginger 10g, dried chili 10g, stir-fried to balance the ingredients then add salt 10g, sugar 10g, stir them well, add green and red bell pepper and onion to continue stir-frying. Sprinkle the minced peanut 50g, cumin 5g, stir well and the dressing is ready.

5. Put the dressing with the chicken strips and well stirred. The dish is done.

中国大锅菜

蒸烤箱卷（纪念版）

The Big-Wok-Made Cuisine of China, Food Volume of Steam Oven（Commemorative Edition）

五花肉烧豆泡

西芹杏鲍菇

菜品特点

特色 麻辣鸡条是一道香辣可口的川菜，由鸡肉、辣椒和洋葱等辅料制作而成。川菜以辣著称，麻辣鸡条也不例外，受到众多吃辣人士的喜欢。

品味 鸡肉通过万能蒸烤箱烹制，软嫩入味，又与红辣椒爆炒，辣味深入肉质的每个地方，吃起来香辣可口、余味萦绕。而洋葱则微甜，进一步提升了鲜香的程度，使菜的滋味更加厚重。

品相 麻辣鸡条这道菜鲜艳红润，红色的辣椒与金黄色的鸡肉互相辉映，辅以青、红椒，给人以视觉上的冲击，菜色丰富，十分诱人。

营养价值 鸡肉是蛋白质最高的肉类之一，是属于高蛋白、低脂肪的食品，且易被人体吸收利用。此外，鸡肉还含有脂肪、钙、磷、铁、镁、钾、钠、维生素等营养成分。

菜品名称

箱子豆腐

— ❧ —

Name: Packed Tofu

制作人：赵春源　　中国烹饪大师

Made by: Chunyuan Zhao　　A Great Master of Chinese Cuisine

主 料 Main Ingredient
豆 腐：1500g 切块
Tofu　1500g　Cut

配 料 Burdening
猪五花：250g 切末
Pork Belly　250g　Minced
鲜 虾：250g 切末
Shrimp　250g　Minced
韭 菜：250g 切末
Leek　250g　Minced
香 菇：100g 切末

Xianggu Mushroom　100g　Minced

小 料 Other Seasoning
姜：5g 切末
Ginger　5g　Minced
葱：5g 切末
Scallion　5g　Minced

调 料 Seasoning
清 油 Oil..........................30ml
盐 Salt.................................20g
酱 油 Soy Sauce 8ml

胡椒粉 Ground Pepper...........8g
鸡 粉 Chicken Powder 5g
淀 粉 Starch......................20g

备 注 Tips
豆腐要烤成金黄色，顶面刷油即可，
油会沿着豆腐的四周流下。
Bake the tofu till golden yellow, only
brush oil on the surface of the tofu
wil befine.

中国大锅菜

The Big-Wok-Made Cuisine of China, Food Volume of Steam Oven（Commemorative Edition）

蒸烤箱卷（纪念版）

制作方法

❶ 首先烤制豆腐，烤盘内刷底油，将切成块的豆腐摆入烤盘，豆腐上面再刷一层油，放入万能蒸烤箱烤制，选择『单点分层炙烤』模式，时长10分钟。

❷ 将肉末、虾末、韭菜末、香菇末加葱末5g、姜末5g、酱油5ml、盐10g、胡椒粉3g，拌成三鲜肉馅。将淀粉20g制成水淀粉备用。

❸ 将烤制好的豆腐块切开一面作『盖子』，注意不要切断，用小刀将豆腐块中间的豆腐肉掏出，这样将豆腐制成一个箱子的形状。

❹ 将调制好的馅料填入『箱子』，合上豆腐盖，放入万能蒸烤箱烹制成熟，选择『单点分层蒸煮』模式，时长7分钟。

❺ 在烹制豆腐时炒制料汁，锅热下油30ml，油热加葱、姜末各10g，酱油3ml，倒入水300ml，锅开后加盐10g，鸡粉5g，胡椒粉5g，倒入水淀粉勾芡，料汁即成。

❻ 将料汁浇在蒸熟的豆腐上，菜品即成。

WORKING PROCESS

1. First of all, bake the tofu. Brush oil at the baking pan, place cut tofu into baking pan, then brush oil on the surface of tofu, bake it in the Universal Steam Oven, select "Single Point Stratified Grill" mode for 10 minutes.

2. Mix and stir the minced pork belly, shrimp, leek, mushroom with minced scallion 5g, minced ginger 5g, soy sauce 5ml, salt 10g, ground pepper 3g. Well stir them together into meat stuff. Turn starch 20g into water-starch mash for later use.

3. Cut the tofu into block. And carve the middle part of tofu to dig a hole. Make tofu like a box.

4. Stuff the meat stuff into the box, steam in the Universal Steam Oven, select "Single Point Stratified Steam" mode for 7 minutes.

5. Make the sauce at the same time, pour oil 30ml into a heated wok, add minced scallion 10g and minced ginger 10g, soy sauce 3ml, pour water 300ml, sprinkle salt 10g after the water boiled, then add chicken powder 5g, ground pepper 5g, pour water-starch mash to thicken the sauce. The sauce is done.

6. Water the sauce on the tofu, the dish is done.

中国大锅菜

菜品名称 · 箱子豆腐
Name: Packed Tofu

菜 品 特 点

特色 箱子豆腐是山东省传统的汉族名菜，属于鲁菜分支博山菜的代表菜之一。载有此品，是满汉全席九白宴中的热菜四品之一。清代御膳膳底档其制作方法是将肉馅填入挖空的豆腐块中，将豆腐烧制金黄，像小箱子一样。

品味 箱子豆腐中的肉馅为三鲜口味，肉末、虾仁、韭菜、香菇各显神通，但又不孤军奋战，合在一起后产生一种诱人的鲜味，将馅料中的鲜味发挥得淋漓尽致。豆腐箱子如饺子皮一般将肉馅包裹，但又具有豆腐本身的香味，入口软嫩，鲜香可口。

品相 箱子豆腐中的豆腐块事先经过烤制，呈金黄色，打开『箱子』盖，馅料鲜嫩多汁，香气扑鼻，从观感上给人以很大的诱惑。

营养价值 豆腐营养丰富，含有铁、钙、磷、镁等人体必需的多种微量元素，还含有糖类、植物油和丰富的优质蛋白，素有『植物肉』之美称。有高蛋白，低脂肪，降血压，降血脂，降胆固醇的功效。大豆蛋白属于完全蛋白质，其氨基酸组成比较科学，人体所必需的氨基酸几乎都有，并且十分容易被消化、吸收。而三鲜馅料由于富含虾仁与五花肉，蛋白质含量十分丰富，与豆腐搭配，相得益彰。

干烧鳕鱼块

Name: Dry-Stewed Codfish Fillets

制作人：郑绍武　　中国烹饪大师

Made by: Shaowu Zheng　　A Great Master of Chinese Cuisine

主 料 Main Ingredient
鳕鱼：3000g 切块
Codfish　3000g　Cut

配 料 Burdening
肥 肉：350g 切丁
Fat　350g　Pieced

小 料 Other Seasoning
姜：30g 切丁

Ginger　30g　Pieced
葱：20g 切丁
Scallion　20g　Pieced
蒜：20g 切片
Garlic　20g　Sliced

调 料 Seasoning
清 油 Oil........................120ml
盐 Salt.................................20g
酱 油 Soy Sauce..............25ml

白 糖 Sugar........................80g
料 酒 Cooking Wine........100ml
醋 Vinegar...........................50ml
郫县豆瓣酱 Pixian Chili Bean
　　　　　Paste...............100g
淀 粉 Starch........................30g

中国大锅菜

菜品名称·干烧鳕鱼块

Name: Dry-Stewed Codfish Fillets

制作方法

❶ 首先腌制鱼块，鱼块入盆，倒入料酒 50ml，盐 15g，酱油 5ml，搅拌均匀，腌制 15 分钟。

❷ 将腌制好的鱼块拍粉，放入万能蒸烤箱炸制，烤盘刷底油，将鱼块均匀摆在烤盘上，选择『单点分层炙烤』模式，时长 10 分钟。

❸ 炒制干烧汁，锅热下油 120ml，油热下肥肉丁煸炒，将肉内的水汽煸出，加郫县豆瓣酱 100g，姜丁 30g，蒜片 20g，料酒 50ml，倒入高汤 800ml，烧开后加糖 80g，醋 50ml，盐 5g，酱油 20ml，出锅前加上葱丁 20 克，浇上料汁即成。

❹ 将料汁浇在炸制好的鱼块上，放入万能蒸烤箱烹制入味，选择『单点分层炙烤』模式，时长 5 分钟，即可出锅。

WORKING PROCESS

1. Marinate the fish fillets first. Put the fish fillets into a basin, pour cooking wine 50ml, salt 15g, soy sauce 5ml, well stir and marinate for 15 minutes.

2. Coat the fish with starch, deep-fry into the Universal Steam Oven, brush oil at the bottom of the baking pan, select "Single Point Stratified Grill" mode for 10 minutes.

3. Make the sauce. Pour oil 120ml into a heated wok, stir-fry the fat pieces at heated oil, add Pixian Chili Bean Paste 100g, ginger pieces 30g, sliced garlic 20g, cooking wine 50ml, pour pure soup 800ml. Add sugar 80g while the soup boiled, vinegar 50ml, salt 5g, soy sauce 20ml, sprinkle minced scallion 20g, the sauce is ready.

4. Pour the sauce on the fried fish, bake at the Universal Steam Oven, select "Single Point Stratified Grill" mode for 5 minutes.

中国大锅菜

The Big-Wok-Made Cuisine of China, Food Volume of Steam Oven (Commemorative Edition)

蒸烤箱卷（纪念版）

菜品特点

特色 干烧是一种重要的烹调方法，将炸制好的主料经小火烧制，使其入味，主要用于制作鱼类。鳕鱼是世界上年捕获量最大的鱼类之一，主要产于大西洋，以前中国较少食用。鳕鱼肉质厚实，刺少，以味道鲜美著称，用中国传统烹饪方法制作鳕鱼本身就是一种创新，而用万能蒸烤箱则将这种创新进一步深化。

品味 鳕鱼肉质紧致成型而不柴，辅以干烧料汁，在其本身的鲜味上又增加了复合香味，肥肉丁和郫县豆瓣酱使口味更加厚重，而糖和醋不但去腥提鲜，也丰富了食用时的味觉体验。

品相 干烧的烹调方法会使鱼身上带有酱红色，十分入味，给人以十足的食欲。咬一口，鱼肉的白嫩便露了出来，展示着鱼肉的鲜美。

营养价值 鳕鱼蛋白质含量非常高，而脂肪含量极低，不到0.5%。更重要的是，鳕鱼的肝脏含油量高达45%，同时含有A、D和E等多种维生素。鱼肉中含有丰富的镁元素，对心血管系统有很好的保护作用，有利于预防高血压、心肌梗死等心血管疾病。鳕鱼低脂肪、高蛋白，低胆固醇、高营养，易于被人体吸收，是老少皆宜的营养食品。

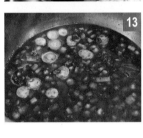

菜品名称

蚂蚁上树

Name: Ants Climbing a Tree

制作人：郑绍武　　中国烹饪大师

Made by: Shaowu Zheng　　A Great Master of Chinese Cuisine

主　料　Main Ingredient
粉　丝：2500g　切段
Vermicelli　2500g　Cut

配　料　Burdening
五花肉：1500g　切末
Pork Belly　1500g　Minced

小　料　Other Seasoning
姜：50g　切片
Ginger　50g　Sliced
葱：30g　切段
Scallion　30g　Cut

调　料　Seasoning
清　油　Oil........................350ml
盐　Salt................................10g
酱　油　Soy Sauce...........100ml
白　糖　Sugar......................10g
料　酒　Cooking Wine.......110ml
郫县豆瓣酱　Pixian Chili Bean
　　　　　　Paste...............200g
花　椒　Chinese Prickly Ash 5g
干辣椒　Dried Chili.................20g

备　注　Tips
1. 一定要将肉末中的水分煸干，否则肉末容易有腥味，没有干香味。
2. 料汁中的高汤要多放，因为粉丝十分吃水。
3. 由于粉丝是白色的，所以要求酱色比较重，最后酱油要根据实际情况选择放多少。

1. Stir-fry the meat till its dry, otherwise the meat is easy to smell, no dry fragrance.
2. Pour more pure soup cause the absorbing of the vermicelli.
3. Balance the color by soy sauce cause the nature color of Vermicelli is white, the amount of soy sauce should be added based on actual situatio.

制作方法

① 首先将粉丝泡入水中备用。此菜可出两布菲芯。

② 煸炒肉末，锅热下油200ml，加肉末煸炒成熟，再加姜末30g，料酒80ml，炒制一会儿，盛出备用。

③ 炒制料汁，锅热下油150ml，油热下郫县豆瓣酱200g，花椒5g，干辣椒20g，姜末20g，葱段30g，料酒30ml，酱油50ml，倒入高汤1.5L，水开加盐10g，白糖10g，用滤网将锅中调料捞出，倒入煸炒好的肉末，再加酱油50ml，料汁即成。

④ 将两个烤盘内刷油，倒入滤水后的粉丝，码放整齐，放入万能蒸烤箱炸制，选择『单点分层煎烤』模式，时长3分钟。

⑤ 将料汁倒入炸制后的粉丝中，放入万能蒸烤箱继续烹制入味，选择『单点分层煎烤』模式，时长4分钟，即可出锅。

WORKING PROCESS

1. Soak the vermicelli by using water.

2. Stir-fry the minced meat, pour oil 200ml into a heated wok, then add minced ginger 30g, cooking wine 80ml. Set aside for later use.

3. Make the sauce, pour oil 150ml into a heated wok, then add Pixian chili bean paste 200g, Chinese prickly ash 5g, dried chili 20g, minced ginger 20g, chopped scallion 30g, cooking wine 30ml, soy sauce 50ml, pour the pure soup 1.5L. Then while the soup boiled, add salt 10g, sugar 10g. Flitter the seasoning, add stir-fry the minced meat, then add soy sauce 50ml, the sauce is ready.

4. Brush the oil at 2 baking pans, cover the vermicelli into the baking pans, bake in the Universal Steam Oven, select "Single Point Stratified Bake" mode for 3 minutes.

5. Pour the sauce into the vermicelli, continue to baking in the Universal Steam Oven, select "Single Point Stratified Bake" mode for 4 minutes.

中国大锅菜

菜品名称·蚂蚁上树
Name: Ants Climbing a Tree

菜 品 特 点

特色 『蚂蚁上树』是一道川菜名菜，其实就是肉末粉丝，炒香的肉末黏在晶莹剔透的粉丝上面，活像密密的蚂蚁爬上树梢。这道菜的典故来源于元代著名戏剧家关汉卿笔下的人物窦娥，她婆婆病重，而家中只剩下一小块猪肉，她便创作了此菜，端上桌时，婆婆误以为此菜上有很多蚂蚁，于是就起了这个名字。

品味 肉末经过煸炒之后干香可口，郫县豆瓣酱炒制的料汁则让此菜的川味十足，麻辣鲜香。粉丝经过泡制后十分软嫩，吸取了料汁后，非常入味。

品相 肉末贴在粉丝上，蚂蚁为肉末，树为粉丝，菜品以形取名，形象逼真。料汁经过红油和酱油炒制，呈暗红之色，与粉丝炒制，粉丝油亮、晶莹。

营养价值 粉丝富含碳水化合物、膳食纤维、蛋白质、烟酸和钙、镁、铁、钾、磷、钠等矿物质。粉丝有良好的附味性，它能吸收各种鲜美汤料的味道，再加上粉丝本身的柔润嫩滑，更加爽口宜人。

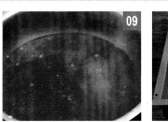

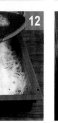

菜品名称

水煮鱼片

Name: Braised Sliced Fish

制作人：郑绍武　　中国烹饪大师

Made by: Shaowu Zheng　　A Great Master of Chinese Cuisine

主　料　Main Ingredient
龙俐鱼：1500g　切片
Long Li Fish　1500g　Sliced

配　料　Burdening
白　菜：1750g　切片
Chinese Cabbage　1750g　Sliced
南　瓜：750g　切片
Pumpkin　750g　Sliced

小　料　Other Seasoning
姜：10g　切片
Ginger　10g　Sliced
葱：40g　切段、切末

Scallion　40g　Cut, Minced
蒜：30g　切片、切末
Garlic　30g　Sliced, Minced

调　料　Seasoning
清　油　Oil.......................300ml
盐　Salt 40g
酱　油　Soy sauce..............15ml
白　糖　Sugar.......................10g
料　酒　Cooking Wine.......110ml
郫县豆瓣酱　Pixian Chili Bean
　　　　　　　Paste..............100g
花　椒　Chinese Prickly Ash ..10g
干辣椒　Dried Chili 20g

辣椒面　Ground Chili..............20g
淀　粉　Starch.......................80g

备　注　Tips
1．腌制鱼时要轻轻搅拌，防止鱼片破碎。
2．烧制料汁时，放入水后要大火开一段时间，使调料的味道充分进入汁中。
1. Slightly stir during the marinating to avoid to break the fish slices.
2. During making the sauce make sure to mix all ingredients to balance the taste.

中国大锅菜

菜品名称·水煮鱼片
Name: Braised Sliced Fish

制作方法

❶ 首先将龙俐鱼片腌制挂糊，将鱼片加盐30g，使鱼肉内的水分出来一些，虑掉多余的水分。将鱼片入盆，加糖10g，料酒10㎖，淀粉80g，油50㎖，轻轻搅拌，使鱼挂浆上糊。

❷ 将白菜片和南瓜片放入漏眼蒸盘飞水，选择『蒸制蔬菜』模式，时长3分钟。

❸ 烧制水煮汁，锅热下油150㎖，油热加花椒10g，干辣椒20g，郫县豆瓣酱100g，蒜片10g，姜片10g，葱段20g，煸炒出香味，倒入料酒100㎖，高汤600㎖，大火烧开，用滤网捞出花椒、干辣椒等，加盐10g，酱油15㎖，料汁即成。

❹ 将飞水后的白菜和南瓜码入盘中，铺上腌制好的鱼片，浇上料汁，放入万能蒸烤箱烹制成熟，选择『单点分层煎烤』模式，时长5分钟。

❺ 出锅后，在菜品表面撒上蒜末20g，葱末20g，辣椒面20g，泼上热油100㎖，即可盛盘。

WORKING PROCESS

1. Coat the Long Li fish with paste. Salt the sliced fish till dehydration, flitter out the excess water, put the fish into basin, add sugar 10g, cooking wine 10ml, starch 80g, oil 50ml, stir slightly to coat the fish.

2. Steam the Chinese cabbage and pumpkin in leaking baking pan, select "Steam- Vegetable" mode for 3 minutes.

3. Stew the sauce. Pour oil 150ml into a heated wok, then add Chinese prickly ash 10g, dried chili 20g, Pixian chili bean paste 100g, sliced garlic 10g, sliced ginger 10g, cut scallion 20g, stir-fry till fragrance, pour cooking wine 100ml, pure soup 600ml, heat till boiling, flitter the seasonings and add salt 10g, soy sauce 15ml, the sauce is ready.

4. Put the steamed Chinese cabbage and pumpkin into baking pan, cover it with marinated fish, pour the sauce, bake in the Universal Steam Oven, select "Single Point Stratified Grill" mode for 5 minutes.

5. Sprinkle the minced garlic 20g, scallion 20g, ground chili 20g. pour the hot oil 100ml, then the dish done.

中国大锅菜

蒸烤箱卷（纪念版）

The Big-Wok-Made Cuisine of China, Food Volume of Steam Oven（Commemorative Edition）

菜 品 特 点

特色 这是一道川菜名菜，起源于重庆地区，『水煮』是著名的川菜烹调方法，虽名为水煮，却看上去很油，但实际上又不是那么油腻，油只是浮在水面而已。

品味 水煮鱼片突出了川菜麻辣鲜香的特点，素有『麻上头，辣过瘾』的说法，红辣椒的辣味不仅遍及汤汁，而且也渗透到鱼肉的每一个缝隙，和鱼肉的鲜嫩交融为一体。那浓重的麻辣口味，总能引诱和刺激人的味觉神经，使人越吃越上瘾、欲罢不能。

品相 水煮汁呈亮红之色，红色的辣椒密集分布，鲜艳耀眼，而嫩白、透亮的鱼肉片参差可见，愈发诱人。

营养价值 此菜鱼肉的选用范围十分广，可以是江河中鱼类，亦可是海鱼，鱼肉低脂肪而高蛋白，味道鲜美，营养价值丰富。这道菜十分适合冬日食用，麻辣的味道，较高的热量，有助于帮助抵御寒冷，荤素搭配的吃法又补充了必要的维生素和植物纤维。

鱼香虾肉

Name: Braised Shrimp

制作人：郑绍武　　中国烹饪大师

Made by: Shaowu Zheng　　A Great Master of Chinese Cuisine

主 料 Main Ingredient
虾 仁：2000g　去虾线
Shrimp　2000g　Cleaned

配 料 Burdening
苦 瓜：1000g　切片
Bitter Gourd　1000g　Sliced
胡萝卜：1000g　切丁
Carrot　1000g　Pieced

小 料 Other Seasoning
姜：50g　切末
Ginger　50g　Minced
葱：80g　切末
Scallion　80g　Minced
蒜：50g　切末
Garlic　50g　Minced

调 料 Seasoning
清 油 Oil........................150ml

盐 Salt 65g
酱 油 Soy Sauce 30ml
白 糖 Sugar........................85g
料 酒 Cooking Wine 90ml
泡 椒 Pickled Pepper 150g
淀 粉 Starch..................350ml
醋 Vinegar.........................150ml
鸡 蛋 Egg................4 个 (pcs)

中国大锅菜

蒸烤箱卷（纪念版）

The Big-Wok-Made Cuisine of China, Food Volume of Steam Oven（Commemorative Edition）

制作方法

❶ 首先腌制虾仁，将30g盐加入虾仁中使虾肉内水分挤出来，滤掉多余水分，盐15g，糖5g，料酒30ml，淀粉30g，腌制15分钟。

❷ 制糊：盆内打入4个鸡蛋打散，加300g淀粉搅拌成糊。将50g淀粉制成水淀粉备用。

❸ 将腌制好的虾仁挂糊，放入万能蒸烤箱进行滑油，选择『鱼类－裹面品－薄』模式，时长4分钟。将苦瓜和胡萝卜放入万能蒸烤箱飞水，选择『蒸制蔬菜』模式，苦瓜时长1分钟，胡萝卜时长5分钟。

❹ 烧制鱼香汁，锅热下油150ml，油热下泡椒碎150g，葱、姜、蒜末各50g，料酒60ml，白糖80g，醋150ml，酱油30ml，盐20g，味精30g，高汤600ml，锅开后倒入水淀粉，撒上葱花30g，料汁即成。

❺ 将滑油后的虾仁和飞水后的胡萝卜、苦瓜盛入布菲盒中，浇上烧制好的鱼香汁，搅拌均匀，即可盛盘。

WORKING PROCESS

1. First marinated shrimps, add salt 30g to the shrimp meat to dehydration, filter out excess water, add salt 15g, sugar 5g, cooking wine 30ml, starch 30g, mix to marinade for 15 minutes.

2. Make paste, add 4 eggs into a basin, knead it with starch 300g till to paste. Turn starch 50g into water-starch mash for later use.

3. Coat the shrimp with paste, bake in the Universal Steam Oven, select "Fish-Coat Powder-Thin" mode for 4 minutes. Steam Bitter Gourd and carrot via the Universal Steam Oven, select "Steam-Vegetable" mode, 1 minute for Bitter Gourd, 5 minutes for carrot.

4. Make the fish-flavor sauce, pour oil 150ml into a heated wok, then add pickled pepper 150g, minced scallion 50g, minced ginger 50g, minced garlic 50g, cooking wine 60ml, sugar 80g, vinegar 150ml, soy sauce 30ml, salt 20g, MSG 30g, pure soup 600ml till boiled. Then add water-starch mash and sprinkle minced scallion 30g, the sauce is done.

5. Pour the baked shrimp, steamed carrot and bitter gourd into buffet pot, pour the fish-flavor sauce and balance all the ingredients. The dish is done.

中国大锅菜

菜品名称·鱼香虾肉

Name: Braised Shrimp

菜品特点

特色 这道菜由传统川菜名菜『鱼香肉丝』演变而来，配料相应改为苦瓜与胡萝卜。与虾肉相比，猪里脊脂肪含量较低，并具有海鲜独特的鲜美之味，口感更加软嫩，苦瓜清爽，胡萝卜微甜，与这道菜的味觉搭配十分合适。

品味 『鱼香』味为川菜中独特的味型，辣度较轻，酸甜适口，透过咸甜酸辣，又萦绕着阵阵姜蒜的香味，形成了典型的符合味道。上浆过的虾肉经过滑油后具有一层面糊，充分吸收鱼香汁的味道，咬一口，虾肉的鲜嫩、料汁的香气让人欲罢不能，而翠绿的苦瓜又能调节油腻，搭配得当。

品相 此菜色泽红润，红中亦有苦瓜的青绿。

营养价值 此菜热量较高，酸甜辣度适口，十分下饭，是补充能量的上佳之选。虾肉营养价值丰富，性温，富含蛋白质，而脂肪含量较低，易于消化，是滋补的佳品。苦瓜中含有丰富的维生素C和苦味武、苦味素。苦瓜素被誉为『脂肪杀手』，能使捏取脂肪和多糖减少。中医上讲苦瓜具有清热消暑、养血益气、补肾健脾、滋肝明目的功效。苦瓜与虾肉相互中和，又各自发挥着营养价值。

菜品名称

宫保虾仁

Name: Kung Pao Shrimp

制作人：郑秀生　　中国烹饪大师

Made by: Xiusheng Zheng　　A Great Master of Chinese Cuisine

主　料 Main Ingredient
虾　仁：2000g　去虾线
Shrimp　2000g　Cleaned

配　料 Burdening
黄　瓜：750g　切片
Cucumber　750g　Sliced
胡萝卜：750g　切片
Carrot　750g　Sliced
花生豆：500g
Peanut　500g

小　料 Other Seasoning
姜：10g　切末
Ginger　10g　Minced
葱：20g　切末
Scallion　20g　Minced
蒜：10g　切末
Garlic　10g　Minced

调　料 Seasoning
清　油 Oil 130ml
盐 Salt 35g

料　酒 Cooking Wine 30ml
干辣椒 Dried Chili 20g
花　椒 Sichuan Pepper 5g
白　糖 Sugar.......................20g
醋 Vinegar...........................50ml
淀　粉 Starch......................65g
酱　油 Soy Sauce 30ml
老　抽 Dark Soy Sauce.......5ml
鸡　蛋 Egg................1 个 (pcs)

中国大锅菜

菜品名称 · 宫保虾仁

Name: Kung Pao Shrimp

制作方法

❶ 首先将虾仁腌制上浆，将一个鸡蛋的蛋清打入虾仁中，加盐5g，淀粉15g，淀粉要分次加入，边搅拌边加，拌匀后腌制15分钟。将50g淀粉制成水淀粉备用。

❷ 腌制时炒制料汁，锅热下油130ml，油热下花椒5g，炒香后捞出，干辣椒20g炒香，干辣椒炒成枣红色后捞出晾凉备用，加入葱花20g、姜末10g、蒜末10g炒香，加料酒30ml，酱油30ml，醋50ml，水400ml，烧开后放入盐30g，白糖20g，老抽5ml，烧开后加水淀粉勾芡，芡汁要稍微稠一些，料汁便制成。

❸ 将腌制上浆后的虾仁滑油，烤盘内刷底油，将虾仁摆入，上面再刷一层油，选择『单点分层煎烤』模式，时长两分钟。将黄瓜片和胡萝卜片放入万能蒸烤箱飞水，胡萝卜时长5分钟，黄瓜时长两分钟。

❹ 将黄瓜和胡萝卜倒入虾仁中，浇上料汁拌匀，再次放入万能蒸烤箱烹制，选择『单点分层煎烤』模式，时长两分钟。

❺ 出锅后，撒上花生米和干辣椒段，即可盛盘。

WORKING PROCESS

1. First marinate the shrimp. Add one egg white into the shrimp, and then add salt 5g, starch 15g. The starch should be added gradually while stirring. After mixing well, marinate for 15 minutes. Mix starch 50g with water, turn it into water-starch mash for later use.

2. Make the dressing at the same time. Heat oil 130ml in the hot wok, add pepper 5g, stir-fried and takes it out till fragrant. Stir-fry dried chili 20g till fragrant, and takes it out when its color turns date red. Cool it down for later use. Add chopped scallion 20g, minced ginger 10g, minced garlic 10g (well stir-fried in the mean while), cooking wine 30ml, soy sauce 30ml, vinegar 50ml and water 400ml. When the water is boiled, salt 30g, sugar 20g, and dark soy sauce 5ml. When the soup is boiled again, thicken the dressing with water-starch mash. When the dressing becomes thicker then it is ready.

3. Slightly stir-fried the marinated shrimp, and then brush oil over the baking pan. Put the marinated shrimp on the pan, brush oil over the shrimp, and select the "Single Point Stratified Grill" mode, grill for 2 minutes. Put cucumber and carrot into the Universal Steam Oven for blanching (5 minutes for carrots and 2 minutes for cucumbers).

4. Put cucumbers and carrots into the shrimp pan, pour the dressing on, mix well, and put back into the Universal Steam Oven. Select the "Single Point Stratified Grill" mode for 2 minutes.

5. When the shrimp is ready, sprinkle with peanuts and dried chili, and then it's ready for the dish.

中国大锅菜

蒸烤箱卷（纪念版）

The Big-Wok-Made Cuisine of China, Food Volume of Steam Oven（Commemorative Edition）

菜品特点

特色 此菜由川菜名菜『宫保鸡丁』演变而来，虾仁比鸡肉更加软嫩鲜美，营养价值更为丰富。用万能蒸烤箱制作此菜，可减少虾仁滑油时的用油量，而黄瓜则将汁液锁在果肉中，更加味美鲜绿。

品味 虾仁鲜香，花生浓香，辅以清脆爽口的黄瓜和胡萝卜，芡汁酸甜可口、红而不辣、辣而不猛、香辣味浓、味道浓厚，芡汁将各种食材的味道统一在一起，而不掩每一种食材的本味。

品相 此菜芡汁浓稠，虾仁白嫩透红，胡萝卜黄中带红，配以鲜绿色的黄瓜，辅以红色的辣椒和金黄的花生，色泽艳丽，而金黄色的宫保汁则使此菜更加色中有味。

营养价值 此菜主辅料种类较多，营养价值十分丰富，虾肉高蛋白而低脂肪，花生油脂丰富，饱含蛋白，胡萝卜和黄瓜则饱含维生素和胡萝卜素，相比宫保鸡丁，此菜脂肪含量较低，更受健美人士的青睐。

菜品名称

大蒜烧鮰鱼

Name: Braised Channel Catfish With Garlic

制作人：郑秀生　中国烹饪大师

Made by: Xiusheng Zheng　A Great Master of Chinese Cuisine

主 料 Main Ingredient	调 料 Seasoning	淀 粉 Starch......................60g

主 料 Main Ingredient
鮰 鱼：4000g　切段
Channel Catfish　4000g　Cut

配 料 Burdening
肥肉片：150g　切片
Sliced Pork Fat　150g　Sliced

小 料 Other Seasoning
姜：120g　切片
Ginger　120g　Sliced
葱：50g　切段、末

葱 Scallion　50g　Cut、Minced
蒜：150g　整瓣
Garlic　150g　The Original

调 料 Seasoning
清 油 Oil........................300ml
盐 Salt...................................45g
酱 油 Soy Sauce..............30ml
干辣椒 Dried Chili.................15g
醋 Vinegar...........................70ml
料 酒 Cooking Wine.........50ml
白 糖 Sugar.........................20g

淀 粉 Starch......................60g
花 椒 Sichuan Pepper..........3g
大 料 Aniseed......................5g
桂 皮 Cassia.........................5g
老 抽 Dark Soy Sauce........5ml

备 注 Tips
肥肉片要煸出油来，这样烧出的鱼会更香。
The chopped pork fat should be stir-fried till oil comes out, so that the fish will be tastier.

中国大锅菜

蒸烤箱卷（纪念版）

The Big-Wok-Made Cuisine of China, Food Volume of Steam Oven (Commemorative Edition)

制作方法

❶ 首先腌制鱼段，将鱼段放入盆中，加盐 20g，料酒 20ml，葱段 20g，姜片 20g，淀粉 30g，搅拌均匀，腌制 15 分钟。将 30g 淀粉制成水淀粉备用。

❷ 在烤盘内刷油，将腌制好的鱼段摆入，鱼上再刷一层油，放入万能蒸烤箱煎制，选择『海鲜－高温炙烤－4 号色』，时长 8 分钟。

❸ 煎鱼时炒制料汁，锅热下油 250ml，油热下大蒜 150g 炸香，放入姜片 100g，葱花 30g 炒香，然后放入肥肉片，大火煸出油，放入干辣椒 15g，花椒 3g，大料 5g，桂皮 5g，料酒 30ml，老抽 5ml，酱油 30ml，醋 70ml，盐 25g，白糖 20g，倒入水 350ml，水开后倒入水淀粉勾芡，料汁即成。

❹ 将料汁浇在飞水后的鱼上，放入万能蒸烤箱再蒸一会儿，选择『单点分层煎烤』模式，时长 5 分钟，出锅后即可盛盘。

WORKING PROCESS

1. Marinate the fish firstly. Put the fish slices into a pot, marinate it with salt 20g, cooking wine 20ml, scallion 20g, ginger 20g, starch 30g for 15 minutes. Turn starch 30g into water-starch mash. Then set aside for later use.

2. Brush oil both at baking pan and fish, place into Universal Steam Oven, select "Seafood-High Temperature Grill color No.4", grill for 8 minutes.

3. Dressing should be done before baking the dish. Heat oil 250ml in a hot wok, stirred fried garlic afterwards, then put into ginger 100g and minced scallion 30g till fragrance. Then put into the chopped pork, stir-fried till the fat comes out. Add dried pepper 15g, pepper 3g, aniseed 5g, cassia 5g, cooking wine 30ml, dark soy sauce 5ml, soy sauce 30ml, vinegar 70ml, salt 25g, sugar 20g, pour water 350ml. Pour water-starch mash after the water boiling. And then the dressing is done.

4. Dressing the blanched fish, then put the fish back into the Universal Steam Oven and then steamed for a while, select "Single Point Stratified Grill" mode for 5 minutes, then get out of the baking pan and finish the dish.

中国大锅菜

菜品名称·大蒜烧鲫鱼
Name: Braised Channel Catfish With Garlic

菜品特点

特色 鲫鱼广泛分布于我国东部各大水系，以长江水系为主，在不同地方有不同叫法，江南称鲫鱼，四川称江团，是长江流域广大人民餐桌上的常客，此菜是一道普通家常菜，深受百姓欢迎。

品味 鲫鱼为江河中之鱼，肉质软嫩鲜香，而大蒜和肥肉片则使鱼的味道更加浓厚。加之葱、姜、花椒、桂皮、大料等众多调味品，使得菜品的味型更加丰富，但各种味道又都无法掩盖鱼肉本身的鲜香。

品相 鱼肉经过炸制略有金黄色，芡汁中调料丰富，浇在鱼段上使鱼肉更加透亮，一瓣瓣大蒜更是味道的写照，只看图像就能感受到浓郁的香气，翻开鱼皮，下面的鱼肉雪白细嫩，让人垂涎欲滴。

营养价值 鲫鱼高蛋白、低脂肪，富含多种维生素和微量元素，是滋补营养佳品；富含生物小分子胶原蛋白质，具有延缓衰老、美容之功效。

金瓜百合

菜品名称

清炖狮子头

Name: Stewed Large Meatballs

制作人：郑秀生　　中国烹饪大师

Made by: Xiusheng Zheng　　A Great Master of Chinese Cuisine

主 料 Main Ingredient
猪五花：3500g 切丁
Pork Belly　3500g　Pieced

配 料 Burdening
荸 荠：500g 切小丁
Chufa　500g　Mini-pieced

小 料 Other Seasoning
姜：35g 切段
Ginger　35g　Cut

葱：35g 切片
Scallion　35g　Sliced

调 料 Seasoning
黄 酒 Shaoxing Wine........35ml
盐 Salt 30g
淀 粉 Starch.......................30g
胡椒粉 Ground Pepper 3g

备 注 Tips
1. 将狮子头余熟定型时，可用小勺

试摁，如果肉丸挺住，说明已经定型。
2. 最后加的蔬菜，冬季加大白菜为宜，夏季可以加小油菜等青菜。
1. Test the shape of meatball during the steaming, you can press the meatball with the spoon and identify the shape.
2. The cabbage or small rape are both fine to mix with the meatball.

中国大锅菜

菜品名称·清炖狮子头
Name: Stewed Large Meatballs

制作方法

① 将葱片 15g，姜片 15g 放入水中泡制，制成葱姜水备用。

② 首先进行腌制处理，将五花肉切成石榴粒大小的肉丁，放入盆中，盐 10g，淀粉 30g，黄酒 15ml，加入葱姜水 100ml，要分多次加入，一边加水一边进行搅拌，搅拌过程中可适当摔打，给肉上劲，搅拌均匀上劲后将荸荠倒入，继续搅拌均匀，腌制 15 分钟。

③ 烧开一锅水备用。将腌制好的肉馅制成每个 70g 左右重量的狮子头，放入锅中余熟定型。

④ 取一烤盘，倒入开水，水中加姜片 20g，葱段 20g，盐 20g，黄酒 20ml，胡椒粉 3g，将定型好的狮子头摆入烤盘中，放入万能蒸烤箱蒸制，选择『蒸制蔬菜』模式，时长 60 分钟。

⑤ 将大白菜放入烤盘，垫在狮子头下面，再放入万能蒸烤箱蒸制 3 分钟，即可出锅。

WORKING PROCESS

1. Brew the sliced scallion 15g and sliced ginger 15g in water. Set aside for alternate.

2. First of all, cut the pork belly into small pieces, mix with salt 10g, starch 30g, Shaoxing wine 15ml, pour into the scallion and ginger water 100ml to stir and beat it well. The last step, put chufa inside. Keep them all together prolongs 15 minutes to make the large meatball.

3. Boil a wok of water. Set aside. Make large meatballs, each ball is about 70g weight. Boil the meatball in water till the shape of the meatball can be fixed.

4. Get a baking pan, pour into the boiling water with sliced ginger 20g, cut scallion 20g, salt 20g, shaoxing wine 20g, ground pepper 3g. Set the meatballs into the baking pan, and put into the Universal Steam Oven, select "Steam Vegetable" mode, the duration of steaming time is 60 minutes.

5. Put cabbage in the baking pan just underneath the meatballs, put into the Universal Steam Oven again, prolongs 3 minutes. The dish is done.

中国大锅菜

蒸烤箱卷（纪念版）

The Big-Wok-Made Cuisine of China, Food Volume of Steam Oven（Commemorative Edition）

菜品特点

特色 这是一道淮扬菜的名菜，由五花肉丁和马蹄制作而成，配料简单而味道不简单，狮子头软而不散，五花肉肥而不腻，汤鲜而不油，对制作者的手法具有很高的要求，在水中蒸煮的时间一定要长。

品味 狮子头入口柔绵鲜香中而有荸荠的清脆爽口，再喝一口鲜美的汤汁，吃几片翠绿的白菜，顿觉香味走遍全身。

品相 狮子头白里微微透红，点缀以马蹄的白色，配以绿色的青菜，让人见肉而不腻，见素而不寡。

营养价值 五花肉富含脂肪和蛋白，马蹄则含有大量的蛋白、粗纤维、维生素，白菜含有丰富的维生素。在品尝美味的同时又是一道对人十分具有补益作用的菜品。尤其是在冬季，不但能补充能量和维生素，鲜美的汤汁则更添暖意。

菜品名称

三鲜豆腐

Name: Steamed Tofu with Ham, Carrot and Bean

制作人：郑秀生　　中国烹饪大师

Made by: Xiusheng Zheng　　A Great Master of Chinese Cuisine

主 料 Main Ingredient
豆 腐：3000g 切块
Tofu 3000g Pieced

配 料 Burdening
火 腿：500g 切丁
Ham 500g Pieced
胡萝卜：250g 切丁
Carrot 250g Pieced
青 豆：250g 清洗

Green Soya Bean 250g Rinsed

小 料 Other Seasoning
葱：30g 切末
Scallion 30g Minced

调 料 Seasoning
清 油 Oil........................100ml
盐 Salt 20g
淀 粉 Starch....................30g

胡椒粉 Ground Bell Pepper 5g
白 糖 Sugar........................20g

备 注 Tips
蒸制豆腐时要加点盐，盐会让豆腐的蛋白凝固，使豆腐不易碎。
Add some salt during steaming the Tofu, it will help to keep the shape which is not easy to fall apart.

中国大锅菜

蒸烤箱卷（纪念版）

The Big-Wok-Made Cuisine of China, Food Volume of Steam Oven（Commemorative Edition）

制作方法

❶ 将豆腐放入万能蒸烤箱进行飞水处理，飞水时豆腐放入一点儿盐，约 10g，选择『蒸制蔬菜』模式，时长 8 分钟。将 30g 淀粉制成水淀粉备用。

❷ 在飞水时炒制料汁，锅热下油 100ml，加葱花 30g，炒香后倒入开水 200ml，加盐 10g，糖 20g，胡椒粉 5g，水开后，倒入水淀粉勾芡，料汁即成。

❸ 将料汁均匀倒入飞水后的豆腐中，撒上火腿、胡萝卜、青豆，再次放入万能蒸烤箱进行烹制，选择『蒸制蔬菜』模式，时长 3 分钟。

WORKING PROCESS

1. Steam the Tofu via the Universal Steam Oven with a little salt, about 10g, select "Steam Vegetables" mode prolongs 8 minutes. Turn the water and starch into water-starch mash for later use.

2. At the same time to make the dressing. Pour oil 100ml into a heated wok, add minced scallion 30g to stir-fried till fragrance then add boiled water 200ml, salt 10g, sugar 20g, ground pepper 5g. Put into water-starch mash to thicken the dressing after the water boiling.

3. Pour the dressing into Tofu evenly. Add ham, carrot, bean. Put into the Universal Steam Oven again to heat, select "Steam Vegetables" mode for 3 minutes.

中国大锅菜

菜品名称·三鲜豆腐

Name: Steamed Tofu with Ham, Carrot and Bean

菜品特点

特色　豆腐起源于中国，已有2100多年的历史，深受我国人民喜爱，中华饮食也围绕豆腐而衍生出大大小小的各色菜肴。这是一道家常养生菜，选用了日常生活中最常见的食材，制作方法简单而味道不简单，豆腐要经过飞水处理，去除其中的苦味，兼有定型的作用。制作此菜，以清淡为要义，清淡中食材的鲜香自现。

品味　豆腐本身含有淡淡的香味，辅以火腿及葱花等各种调味料，使其本味之外更添鲜香，咸鲜适口。

品相　豆腐白嫩，激发食欲，胡萝卜和火腿红亮以增鲜美，点缀以绿色的青豆以增清淡，简单之中而不落俗气。

营养价值　豆腐营养丰富，含有铁、钙、磷、镁等人体必需的多种微量元素，还含有糖类、植物油和丰富的优质蛋白，素有『植物肉』之美称，含有高蛋白、低脂肪。具有降血压、降血脂、降胆固醇的功效。大豆蛋白属于完全蛋白质，其氨基酸组成比较好，人体所必需的氨基酸它几乎都有，并且十分容易被人体消化、吸收。

13

菜品名称

黑玉米枣糕

❦

Name: Date Cake Baked with Black Corn Flour

制作人：俞世清　　中国烹饪大师

Made by: Shiqing Yu　　A Great Master of Chinese Cuisine

主 料 Main Ingredient	牛　奶：500g	红　糖 Brown Sugar...........300g
黑玉米面：2000g	Milk　500g	酵　母 Yeast.........................15g
Black Corn Flour　2000g	红　枣：150g	泡打粉 Baking Powder 15g
	Red Jujuba　150g	蓝　莓 Blueberry 50g
配 料 Burdening		
鸡　蛋：1000g	**调 料 Seasoning**	
Egg　1000g	奶　油 Cream 50g	

中国大锅菜

菜品名称·黑玉米枣糕

Name: Date Cake Baked with Black Corn Flour

制作方法

❶ 将鸡蛋打入盆中，用打蛋器打散，加入牛奶和奶油打匀，将温水化开的酵母和红糖水倒入蛋液中搅散、调匀，制成蛋汁。

❷ 将黑玉米粉分三次放入调匀的蛋汁中，调制稠糊。

❸ 蒸盘中点入糕点纸，刷上少许油，将稠糊倒入，用面刀刮平表面，均匀撒入红枣和蓝莓，用手轻按压实。

❹ 放入万能蒸烤箱蒸制成熟，选择『单点分层蒸煮』模式，时长40分钟。

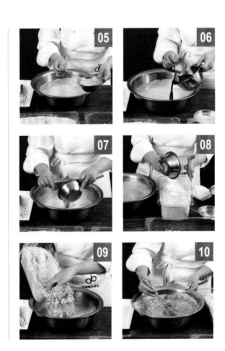

WORKING PROCESS

1. Beat the eggs into a bowl and whisk it with an eggbeater. Add milk and cream and whisk it evenly. Pour the yeast and brown sugar water dissolved in warm water into the egg liquid and whisk it evenly and thoroughly to make the egg juice.

2. Put the black corn flour into the mixed egg juice for three times , and mix it into a paste.

3. Place the cake paper into the steaming tray and brush a little oil, then pour the paste on it and smooth the surface with a knife.Evenly spread red Jujubas and blueberries and press gently with hand to compaction.

4. Put in the Universal Steam Oven to steam. Select "Single Point Stratified Steaming" mode for 40 minutes.

中国大锅菜

蒸烤箱卷（纪念版）

The Big-Wok-Made Cuisine of China, Food Volume of Steam Oven（Commemorative Edition）

菜品特点

特色 枣糕是人们十分喜爱的面点之一，它质地松软，口味香醇，十分适合老人和小孩食用。食材选择上，以沧州地区的金丝小枣为最佳，它核小肉多，含糖量高，是枣中的精品。

营养价值 黑玉米是玉米的一种特殊品种，其籽粒角质层不同程度地沉淀有黑色素，所以外观乌黑发亮。籽粒富含水溶性黑色素、各种人体必需的微量元素、植物蛋白质和各种氨基酸等，营养含量高于其他谷类作物。

菜品名称

红豆烧饼

Name: Red Bean Paste Clay Oven Rolls

制作人：俞世清　　中国烹饪大师

Made by: Shiqing Yu　　A Great Master of Chinese Cuisine

主 料 Main Ingredient	配 料 Burdening
面 粉: 1000g	猪 油: 400g
Flour　1000g	Lard　400g
红豆馅: 1000g	芝 麻: 50g
Red Bean Paste　1000g	Sesame　50g

中国大锅菜

蒸烤箱卷（纪念版）

The Big-Wok-Made Cuisine of China, Food Volume of Steam Oven（Commemorative Edition）

制作方法

❶ 将500g面粉开窝，加入250g猪油，使劲揉搓，制成『油心面』备用。

❷ 将另外500g面粉开窝，将猪油150g置于中间，用少许水化开，加水230ml。和面时要使劲摔面，摔到表面滋润光滑，制成『水油面皮』备用。

❸ 将和好的『水油皮面』取200g包入40g『油心面』，擀制3mm厚，进行折叠，再擀开，然后顺一边卷起成条状，揪成40g左右的剂子，收口向上。将收口对折，摁成饼状，掸上面粉，擀制成面皮。

❹ 将面皮放入手中，包入红豆馅，顺着一个方向包制，收口处收严，蘸上芝麻。将包好的馅饼均匀摆入刷好油的烤盘中进行烤制，选择『面包类－干烤曲奇－5号色』，时长12分钟。

WORKING PROCESS

1. Make a nest with 500g flour, add 250g lard and knead it forcefully to make the "oil hearted dough" for later use.

2. Make a nest for the other 500g flour and place 150g lard in the middle.Melt it with a little water, add water 230ml. When making the dough, smash the dough forcefully till its surface becomes smooth and moist.Make it into "water oil dough" for later use.

3. Take 200g "water oil dough" and wrap it into 40g "oil hearted dough". Roll it into 3mm thick, then fold and roll spread. Roll it into a strip along one side. Break it into pieces of 40g.Close up and fold, press it into the pie.Whisk flour and roll it into wrapper.

4. Wrap the red bean paste with the wrapper and close it along one direction, seal the closing up point tightly and dip in sesame. Place the wrapped pie evenly into the baking tray brushed with oil to bake. Select "Bread and Pastries- Baking Cookies - Color No.5" mode for 12 minutes.

中国大锅菜

菜品名称·红豆烧饼

Name: Red Bean Paste Clay Oven Rolls

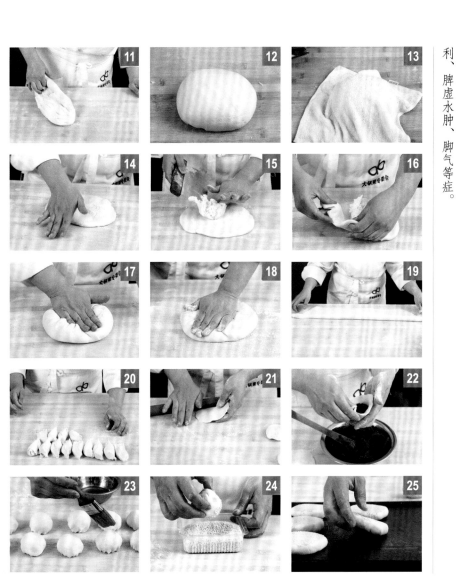

菜品特点

特色 红豆沙馅烧饼是我们小时候的最爱，表皮酥脆，馅料香甜，这份甜蜜久久跟随着我们。红豆十分适合制作馅料，有条件的话，可以购买红豆自己制作馅料，甜度可自己掌控。

营养价值 红豆具有清热解毒、健脾益胃、利尿消肿、通气除烦等功能，可治疗小便不利、脾虚水肿、脚气等症。

菜品名称

萝卜丁焗饭

Name: Steamed Rice with Diced Radish and Carrot

制作人：俞世清　　中国烹饪大师

Made by: Shiqing Yu　　A Great Master of Chinese Cuisine

主　料 Main Ingredient	White Radish　250g　Diced	调　料 Seasoning
大　米：1500g	腊　肉：100g　切丁	清　油 Boiled Oil 100ml
Rice　1500g	Bacon　100g　Diced	盐 Salt 20g
	腊　肠：100g　切丁	酱　油 Soy Sauce 30ml
配　料 Burdening	Sausage　100g　Diced	胡椒粉 Pepper Powder..........5g
胡萝卜：250g　切丁	海　米：50g	葱 Green Onion.....................20g
Carrot　250g　Diced	Dried Shrimps　50g	姜 Ginger.............................20g
白萝卜：250g　切丁		

中国大锅菜

菜品名称・萝卜丁焗饭
Name: Steamed Rice with Diced Radish and Carrot

（制作方法）

❶ 将大米提前泡制 30 分钟备用。

❷ 锅热下油 100ml，油温在三成热时加入腊肉、腊肠和海米，煸炒出香味，加葱、姜末各 20g 炒香，加酱油 30ml 上色，加胡椒粉 5g、盐 20g。

❸ 将炒好的料汁倒入深布菲盒中，倒入大米，拌匀，撒入胡萝卜丁和白萝卜丁，加入适量清水拌匀，选择『单点分层蒸煮』模式，时长 40 分钟，菜品即成。

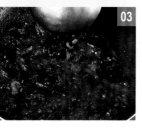

WORKING PROCESS

1. Soak the rice for 30 minutes in advance for later use.

2. Add oil 100ml on the heated pot and add 100g bacon, sausage and dried shrimps after the oil temperature reaches 30% to stir-fry out the fragrance. Add green onion 20g, minced ginger 20g to fry and then add soy sauce 30ml for coloring. Then add pepper powder 5g, salt 20g.

3. Pour the cooked sauce into a deep container and pour in the rice to stir evenly. Add appropriate amount of fresh water to stir evenly. Spread the diced carrots and white radish and put in the Universal Steam Oven. Select "Single Point Stratified Steaming" mode for 40 minutes and the dish is done.

中国大锅菜

蒸烤箱卷（纪念版）

The Big-Wok-Made Cuisine of China, Food Volume of Steam Oven（Commemorative Edition）

菜 品 特 点

特色 萝卜丁焗饭便于制作，味道不俗，米饭充分吸收了腊肉、腊肠中的油脂，变得十分可口，香气浓郁。这道菜类似于广式煲仔饭，只不过要将腊肠、腊肉和海米提前炒制一番，更能上色、入味，而胡萝卜和白萝卜则使菜品的香气更加清香诱人。使用万能蒸烤箱烹制此菜，可以十分精确控制温度和风速，避免出现煳锅现象。

营养价值 大米最主要的成分是碳水化合物约70%，同时含有蛋白质7%至8%、维生素、纤维素和矿物质，是补充营养素的基础食物。

番茄焗面

苹果馅饼

千层酥卷

肉 龙

菜品名称

麦香馅饼

Name: Meat Pie

制作人：俞世清　　中国烹饪大师

Made by: Shiqing Yu　　A Great Master of Chinese Cuisine

主 料 Main Ingredient

面 粉：1000g
Flour　1000g

牛肉末：400g
Minced Beef　400g

配 料 Burdening

猪 油：400g
Lard　400g

洋 葱：250g 切丁
Onion　250g　Diced

调 料 Seasoning

清 油 Boiled Oil 20ml
酱 油 Soy Sauce 10ml
香 油 Sesame Oil 20ml
胡椒粉 Pepper Powder..........3g
葱 Green Onion.....................5g

姜 Ginger...............................5g

备 注 Tips

和好面粉后，表面要盖上湿布，防止表面水分蒸发导致干裂。

After making the dough, cover a wet cloth on top to avoid cracking due to the moisture evaporation of the surface.

中国大锅菜

蒸烤箱卷（纪念版）

制作方法

❶ 将500g面粉开窝，加入250g猪油，使劲揉搓，制成『油心面』备用。

❷ 将另外500g面粉开窝，将猪油150g置于中间，用少许水化开，加水230ml。和面时要使劲摔面，摔到表面滋润光滑，制成『水油面皮』备用。

❸ 制作肉馅，锅热下油20ml，加入100g牛肉馅煸炒，肉末变色后加葱、姜各5g炝锅，加入酱油10ml上色，加胡椒粉3g。将炒好的牛肉末倒入盆中，散去热气后，倒入剩余的300g牛肉馅，加入洋葱丁250g，香油20ml，搅拌均匀，制成馅料。

❹ 将和好的『水油皮面』取200g包入40g『油心面』，擀制3mm厚，进行折叠，再擀开，然后顺一边卷起成条状，揪成40g左右的剂子，收口向上。将收口对折，摁成饼状，掸上面粉，擀制成面皮。

❺ 将面皮平摊在手中，将肉馅填实，顺着一个方向包制，收口处收严。将包好的馅饼均匀摆入刷好油的烤盘中进行烤制，选择『面包类—干烤曲奇—5号色』，时长12分钟。

WORKING PROCESS

1. Make a nest with 500g flour, add 250g lard and knead it forcefully to make the "oil hearted dough" for later use.

2. Make a nest for the other 500g flour and place 150g lard in the middle. Melt it with a little water,then add water 230ml. When making the dough, smash the dough forcefully till its surface becomes smooth and moist. Make it into "water oil dough" for later use.

3. Make the minced meat. Add oil 20ml in the heated pot. Add minced beef 100g and stir-fry till it changes color. Then add green onion, ginger each 5g to stir-fry. Add soy sauce 10ml for coloring and add pepper powder 3g. Pour the cooked minced beef in a basin and wait for the heat to go. Then pour in the remaining 300g minced beef, and add diced onion 250g, seaman oil 20ml to stir evenly to make the filling.

4. Take 200g "water oil dough" and wrap it into 40g "oil hearted dough". Roll it into 3mm thick, then fold and roll spread. Roll it into a strip along one side. Break it into pieces of 40g. Close up and fold, press it into the pie. Whisk flour and roll it into wrapper.

5. Wrap the meat filling with the wrapper and close it along one direction; seal the closing up point. Place the wrapped pie evenly into the baking tray brushed with oil to bake. Select "Bread and Pastries- Baking Cookies - Color No.5" mode for 12 minutes.

中国大锅菜

菜品名称·麦香馅饼

Name: Meat Pie

菜品特点

特色 制作这道面点对手艺要求较高，水、猪油、面均需精确用量，才能保证出层层酥脆的面皮，经过烤制，麦香四溢。肉馅的调制也十分讲究，要先将少部分肉馅进行煸炒，这样馅料中既有生牛肉馅的醇香，又有熟牛肉的干香。虽然制作起来比较麻烦，但这种美味，一定能收获食客满意的笑容。

营养价值 牛肉高蛋白低脂肪，营养价值十分高，氨基酸组成比猪肉更接近人体需要，能提高机体抗病能力，且有补中益气、滋养脾胃、强健筋骨的功效，并能起暖胃的作用，十分适合在冬天食用。白萝卜含丰富的维生素C和微量元素锌，有助于增强机体的免疫功能，提高抗病能力。

小 鸭 酥

Name: Duckling-Shaped Crispy puff

制作人：王素明　　中国烹饪大师

Made by: Suming Wang　　A Great Master of Chinese Cuisine

主 料 Main Ingredient	配 料 Burdening
面 粉：1000g	豆沙馅：750g
Flour　1000g	Red Bean Paste　750g
	猪 油：400g
	Lard　400g

中国大锅菜

菜品名称·小鸭酥

Name: Duckling-Shaped Crispy puff

制作方法

① 将500g面粉开窝，加入250g猪油，使劲揉搓，制成『油心面』备用。

② 将另外500g面粉开窝，将猪油150g置于中间，用少许水化开，加水230ml°和面时要使劲摔面，摔到表面滋润光滑，制成『水油面皮』备用。油心面和水油面皮的软硬程度要一致。

③ 开酥，用水油皮面裹住油心面，擀成长饼状，卷成条，用鸡蛋清封口粘住，揪成一个个剂子，擀成面皮。

④ 面皮中包入豆沙馅，包成球状，摁扁，用刀切开两个口，捏成小鸭状，表面刷一层蛋液。放入万能蒸烤箱烤熟，选择『干烤曲奇－3号色』模式，时长10分钟。

WORKING PROCESS

1. Add 500g flour make a hole, lard 250g, and knead, made "oil heart dough". Set aside for later use.

2. Will open another 500g flour, lard 150g in the middle, with a little hydration open, add water 230ml. When making the dough, smash the dough forcefully till its surface moist and smooth. made it into "water oil dought". Set aside for later use. Oil surface and degree of The degree of hard and soft of "oil heart dough "and "water oil dought" must be the same.

3. Wrap the "dough with lard oil"with "water-oil wrapper", roll into a pie shape and roll up into a strip shape, seal with the egg white, cut out of dough to make small pieces, and roll them to wrappers.

4. Put bean-paste in a dough wrapper, shape it into a ball and flatten it.Cut two openings in the dumpling, knead it into the shape of duck and spread it with egg wash. Put it in the multi-function steaming oven to bake in the "cookie toasting-color 3" mode for 10 minutes.

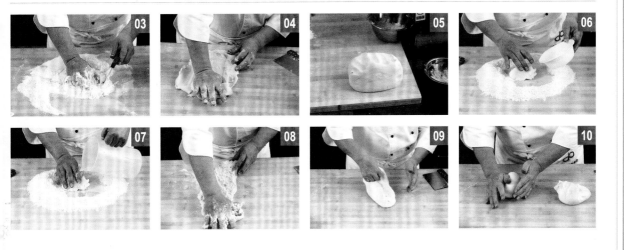

中国大锅菜

The Big-Wok-Made Cuisine of China, Food Volume of Steam Oven（Commemorative Edition）

蒸烤箱卷（纪念版）

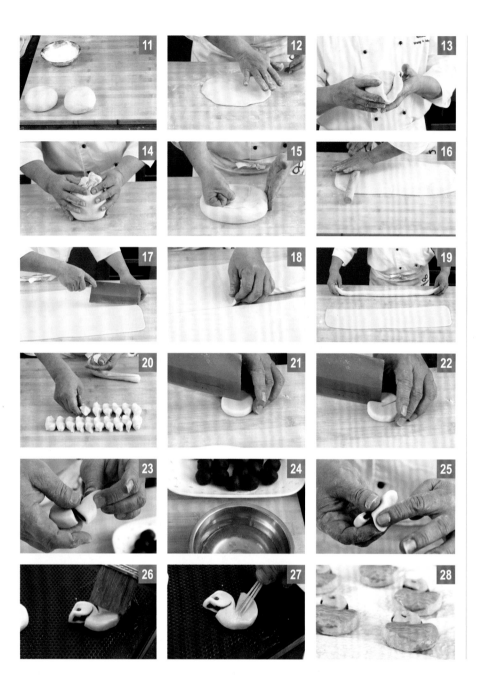

菜 品 特 点

特色 小鸭酥用红豆沙馅制作，外皮酥脆，馅料香甜，造型可爱。

营养价值 红豆有清心养神、健脾益肾功效，中医学认为，红豆有化湿补脾之功效，对脾胃虚弱的人比较适合。红豆最主要的功能就是帮助排水，具有消肿、轻身的功效。

菜品名称

麻酱烧饼

Name: Sesame Crisp Cake

制作人：王素明　　中国烹饪大师

Made by: Suming Wang　　A Great Master of Chinese Cuisine

主 料 Main Ingredient
富强粉：750g
Sesame Crisp pancake flour　750g

配 料 Burdening
芝麻酱：200g
Sesame Sauce　200g

芝 麻：200g
Sesame　200g

调 料 Seasoning
小苏打 Baking Soda...............7g
干酵母 Dry Yeast 1.5g
盐 Salt 9g

花椒面 Pepper.......................7g
小茴香末 Fennel Powder 7g
鸡 蛋 Egg................2 个 (pcs)

中国大锅菜

蒸烤箱卷（纪念版）

The Big-Wok-Made Cuisine of China, Food Volume of Steam Oven（Commemorative Edition）

制作方法

❶ 将面粉开成窝型，小苏打放在中间，倒入水。加入用水化开的酵母，分次加入清水 275ml 和面，活成麦穗型后再加水，这样便于掌握面的软硬。和面时要揉上劲，将面揉透、和匀，然后盖上潮布，醒发 20 分钟。

❷ 将醒发好的面团用压面机压成长方形薄片，面片上刷油，抹上芝麻酱 200g，将花椒面 7g、小茴香末 7g 和盐 9g 混合，制成椒盐，均匀撒在面片上，再撒入少许干面粉。

❸ 将面皮卷起，成条状，切成 70g 左右的剂子，用手揉成饼状，一面刷上蛋液，蘸上芝麻，摆入烤盘，选择『干烤曲奇 -4 号色』，时长 10 分钟。

WORKING PROCESS

1. Add baking soda into the water, melt yeast placed in the middle of the flour, add water 275ml, easy to mastery of the soft and hard surfaces. And when it will pull its weight, surface penetration, and rub evenly, and then covered with a wet towel, for about 20 minutes.

2. Noodle machine pressed into rectangular sheet for the proofing of the dough, brush oil patch, wipe 200g sesame sauce, pepper 7g, fennel powder 7g and salt 9g mixed, made of salt and pepper, evenly sprinkle on the patch, and then thrown into a little dry flour.

3. The dough rolled up, into strips, cut in 70g for each, rubbing by hands into a pie, brush the egg wash on one side, dipped in sesame seeds, put into the pan, select "dry baking cookies- color No.4" for 10 minutes.

中国大锅菜

菜品名称·麻酱烧饼
Name: Sesame Crisp Cake

菜品特点

特色 烧饼是东汉时从西域传进来的，范晔的后汉书记载有：『灵帝好胡饼。』胡饼就是最早的烧饼。胡饼到了唐代更加流行，白居易有诗寄胡饼与杨万州，其中『胡麻饼样学京都，面脆油香新出炉』两句道出了胡饼的美味与流行。烧饼在我国的流传过程中也衍生出了无数的品种，其中麻酱烧饼就是流行北方最为常见的一种。麻酱烧饼外皮酥脆、内筋软，层次分明。花椒盐香味、麻酱香味、芝麻香味交相辉映，干香回味。

营养价值 芝麻酱富含蛋白质、脂肪及多种维生素和矿物质，其中含钙和铁都比较多，有很高的营养价值，经常适量食用对骨骼、牙齿的发育都有益处。此外，由于芝麻含铁量高，不仅对调整偏食厌食有积极的作用，还能纠正和预防缺铁性贫血。

菜品名称

肉丁包子

制作人：王素明　　中国烹饪大师

Name: Diced Meat Steamed Stuffed Bun

Made by: Suming Wang　　A Great Master of Chinese Cuisine

主　料　Main Ingredient
面　粉：1500g
Flour　1500g
五花肉：1000g　切丁
Pork Belly　1000g

配　料　Burdening
豇　豆：1000g　切丁
Cowpea　1000g　Pieced
涨发宽粉：200g　切丁
Wide Vermicelli　200g　Pieced
涨发木耳　150g　切碎
Black Fungus　150g　Minced

葱　白　250g　切丁
Fistular Onion Stalk　250g　Pieced

调　料　Seasoning
酵　母　Yeast..........................18g
泡打粉　Baking Powder..........18g
白　糖　Suger.......................150g
黄　酱　Yellow Soybean.......200g
甜面酱　Sweet Sauce...........100g
胡椒粉　Pepper Powder...........3g
八　角　Anise........................10g
香　油　Sesame Oil..............50ml

熟猪油　Pork Oil...................60ml
料　酒　Cooking Wine.........25ml
葱　花　Chive........................20g
姜　末　Ginger......................20g

备　注　Tips
没有擀成面皮的剂子要用湿布盖好，因为发面表面特别容易起皮，这样是为了保持剂子表面的湿润。
No rolling into a sub-agent face with a wet cloth cover, because the surface increase particularly easy peeling, it is to keep the surface wet.

中国大锅菜

菜品名称·肉丁包子
Name: Diced Meat Steamed Stuffed Bun

制作方法

① 锅热加入60ml熟猪油炒化，加八角10g，炸香后捞出。倒入五花肉丁，煸炒出水分。加入葱丁250g，葱花20g，姜末20g，料酒25ml，翻炒片刻，加入100g甜面酱和200g用水泄开的黄酱，撒入胡椒粉3g，炒出酱香味后倒入香油20ml即可。

② 将豇豆丁和木耳放入万能蒸烤箱蒸熟，选择「单点分层蒸煮」模式，时长10分钟。

③ 酱肉凉透后入盆，加宽粉条、木耳、豇豆丁，撒入葱白丁，倒入香油30ml，拌匀后即制成馅料。

④ 面粉开成窝，加泡打粉18g，面窝内加白糖150g，加水化开，将酵母18g用水化开后撒入面窝，分次加入清水800ml和面，将面团揉紧、揉均匀，手要使上力道。和好面后用湿布盖上，醒发20分钟。

⑤ 将醒发好的面表面撒上少许面粉，用压面机压平，将面皮裹起来成条状，揪成85g的剂子，擀成面皮，包入馅料。

⑥ 将包好的包子放入万能蒸烤箱发酵，选择「面包类－发酵」模式，时长30分钟。随后蒸熟，选择「单点分层蒸煮」模式，时长16分钟。

WORKING PROCESS

1. Add 60ml lard stearin in a hot pot to fry, add 10g anise to fry the fish out, put the diced meats for stir-fry, add 250g diced green onion, 20g chopped green onion, 20g bruised ginger and 25ml cooking wine, stir fry for a moment, add 100g sweet sauce, 200g yellow soybean diluting with water, sprinkle with 3g pepper, fry until the sauce flavor, and then put 20ml sesame oil. The dish is done.

2. Put the diced tacos and agarics into the universal steam oven to steam, and select the mode of "single -point layering cooking" for of 10 minutes.

3. Cool the spiced port then put it into a dish, add the wide vermicelli, agaric, the diced cow pea, sprinkle with the diced scallion, put 30ml sesame oil, mix thoroughly to make stuffings.

4. Make the flour concave, add 18g baking powder. add 150g white sugar in the flour concave, and add water to melt. After melting 18g yeast with water, scatter them in the flour concave, add 800ml water for sevenal times, knead the dough evenly with hands strength. After kneading dough, conduct the final fermentation for 20 minutes.

5. Sprinkling a little flour on the surface of flour after the final fermentation, flatten with a press flour machine, wrap the dough into strips, make 85g small pieces of dough, roll them to wrappers, and put stuffings in them.

6. Put the steamed bun with stuffing into the universal steam oven for fermentation, select the mode "bread-fermentation" for 30 minutes. Then steam it and select the mode "single-point layering cooking" for 16 minutes.

中国大锅菜

蒸烤箱卷（纪念版）

The Big-Wok-Made Cuisine of China, Food Volume of Steam Oven (Commemorative Edition)

菜品特点

特色

酱肉起源于清代康熙年间，一个名落孙山的秀才开了一家熟肉店，又阴差阳错发现了一种使肉更加鲜美肥嫩的方法，并受到康熙皇帝嘉奖。后人用五花肉制成酱肉丁做馅，这就是我们今天的酱肉包。肉丁肥肉浓香而不腻，咬一口，酱香浓郁，肥厚利口。酱肉大包在北方十分流行，让人联想到七尺壮汉。但这粗犷的背后却有着工艺的细腻，切忌绞馅，要手工切制而成，除水后酱烧，用细火慢炖，肉皮软烂，如此一来，肉丁肥肉浓香而不腻，瘦肉有质感。

营养价值

肉丁包子不但好吃，也集各种营养于一身，五花肉既能补充丰富的动物蛋白，又具有补肾养血的功效；豇豆提供了易于消化吸收的优质蛋白质，适量的碳水化合物及多种维生素、微量元素等，可补充机体的各项营养素；木耳则具有益气、润肺、补脑、轻身、凉血、止血、涩肠、活血、养颜等功效。

菜品名称

蛋丝烧卖

—— ❧ ——

Name: Egg Sumai

制作人：王素明　　中国烹饪大师

Made by: Suming Wang　　A Great Master of Chinese Cuisine

主　料　Main Ingredient
雪花粉：250g
Flour　250g

配　料　Burdening
鸡　蛋：200g
Egg　200g
五花肉：1000g　绞馅

Pork Belly　1000g　chopped
荸　荠：250g　切碎
Water Chestnut　250g　Original

调　料　Seasoning
盐　Salt 20g
胡椒粉　Pepper.......................3g
酱　油　Soy Sauce 20ml

白　糖　Suger........................10g
香　油　Sesame oil 20ml
淀　粉　Starch.......................20g
葱　花　Chive 50g
姜　末　Ginger powder..........20g

中国大锅菜

蒸烤箱卷（纪念版）

The Big-Wok-Made Cuisine of China, Food Volume of Steam Oven（Commemorative Edition）

制作方法

❶ 将鸡蛋打散，加淀粉20g，摊成鸡蛋饼，切成蛋丝备用。

❷ 将面粉开成窝型，加入1个鸡蛋的蛋清，分次加入水共90ml和面，揉面要用力，将和好的面团盖上潮布，醒发20分钟。

❸ 荸荠倒入五花肉，加葱花50g，姜末20g，盐20g，胡椒粉3g，酱油20ml，白糖10g，香油20ml，搅拌均匀，制成馅料。

❹ 将面团搓成条状，揪出剂子，用枣核棍擀制成带荷叶褶的面皮，包入馅料，加蛋丝收口。

❺ 将包好的烧卖放入万能蒸烤箱蒸熟，选择『单点分层蒸煮』模式，时长8分钟。

WORKING PROCESS

1. Mix the egg with 20g starch, spread into the egg cake, then thin slice the egg cake.

2. Add 1 egg white in to flour and 90ml water to make the dough, then let the dough covered with a wet towel, proofing for 20 minutes.

3. Add water chestnut into the pork paste, add chopped chive 50g, minced ginger 20g, salt 20g and pepper 30g, soy sauce 20ml, sugar 10g, sesame oil 20ml, stirring evenly.

4. Rubbing the dough into strips, each with a stick by rolling out, made with ruffle dough fillings into the bag, add egg slice.

5. Wrap the steamed pork sumai into universal steam oven, select "Single Layered Cooking Mode" for 8 minutes.

中国大锅菜

菜品名称·蛋丝烧卖

Name: Egg Sumai

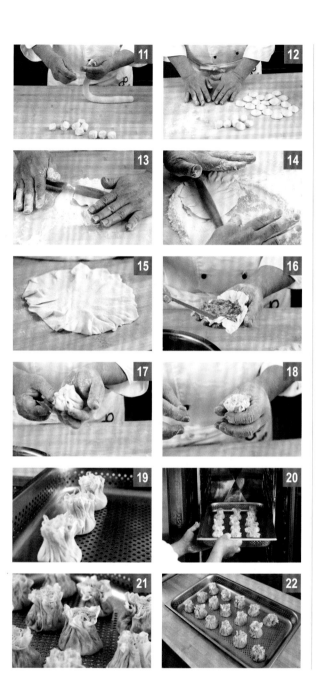

【菜品特点】

【特色】烧卖在我国南北方地区都广为流行。北方以牛、羊肉大葱馅为主，南方则以猪肉或糯米馅为主。根据不同地域，其名称亦有『稍麦』『捎卖』等。它与包子类似，不同的是面皮为烫面，顶端开口，露出馅料。皮薄馅大，鲜香无比。据学者考据，烧卖最早出现于元代，是蒙古人喜爱的美食。最早可能叫作『捎卖』，由于制作简单，就是『捎带着卖』的意思。如果在北方饭馆点烧卖，其重量是按照皮来算，一两是8个左右。

【营养价值】五花肉富含优质的动物蛋白，有补肾、养血、润燥之功效；荸荠则既能当蔬菜，又能当水果，是一种不可多得的两用食物，含有大量的蛋白质、脂肪、粗纤维、胡萝卜素、维生素B、维生素C、铁、钙、磷和碳水化合物，有预防急性传染病之功效。

菜品名称

江米凉果

Name: Glutinous Rice Candied Fruit

制作人：王志强　　中国烹饪大师

Made by: Zhiqiang Wang　　A Great Master of Chinese Cuisine

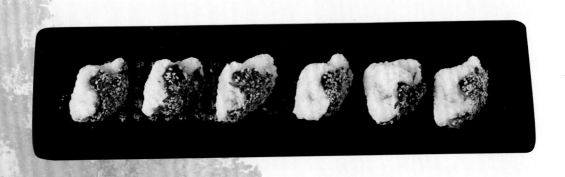

主　料 Main Ingredient
江　米：500g
Glutinous rice　500g

配　料 Burdening
豆　沙：250g

Red bean paste　250g
面　粉：1000g
Flour　1000g

调　料 Seasoning
熟芝麻 Cooked Sesame 10g
白　糖 Sugar.......................10g

青红丝 Green Red Silk..........10g

备　注 Tips
面粉以低筋面粉为好。
It is better to choose the soft flour

中国大锅菜

制作方法

❶ 首先将江米 500g 加水 450ml，放入万能蒸烤箱蒸熟，选择『单点分层蒸煮』模式，时长 40 分钟。将面粉放入万能蒸烤箱干蒸，选择『单点分层蒸煮』模式，时长 40 分钟。

❷ 将熟芝麻、白糖、青红丝各 10g 放入碗中，拌成蘸料备用。

❸ 将蒸制好的江米用麻布揉软，板底和布菲盒中倒入蒸熟的面粉做薄面，江米擀成条，揪成一个个的剂子。

❹ 将剂子摁扁，抹上豆沙馅，扣过来蘸上蘸料，用 3 个手指头捏住拿出，摆入盘中，这道面点就做好了。

WORKING PROCESS

1. First add 500g glutinous rice and 450ml water into the universal steam oven, select "Single Layered Cooking" mode for 40 minutes. Put the flour into the universal steam oven to dry up, select "single point stratified steam" mode for 40 minutes.

2. Add cooked sesame 10g, sugar 10g, green red silk 10g into the bowl to mix up evenly.

3. Crumple steamed glutinous rice prepared with sackcloth knead until soft, bottom plate and buffet pour steamed flour to make a thin face, roll glutinous rice into strips, pulling into one each.

4. Pressed each small dough flat, wipe a red bean paste, an upside down dip the dips, with three finger pinch out, put into the dish, the pasta is done.

中国大锅菜

蒸烤箱卷（纪念版）

The Big-Wok-Made Cuisine of China, Food Volume of Steam Oven（Commemorative Edition）

菜品特点

特色　江米凉果是津门地区的称呼，老北京称之为艾窝窝，是著名的清真风味小吃。食之，凉丝丝、甜蜜蜜，混合着豆馅的香味，直沁肚腹，消夏解暑。衍变到今天，制作这道面点，其中的馅料已更加丰富，但最为经典的还是豆沙馅。江米凉果因食材与口感而得名，不过南方的朋友千万不要误解，此凉果非彼凉果，因为『凉果』一词在南方指腌制的鲜果类。

营养价值　江米被称为谷类中的白珍珠，因为江米的营养价值非常高，它含有较多脂肪、蛋白质和碳水化合物，当然还含有很多的矿物质和维生素，对人体不仅有养生的功效，还具有美容的作用。江米含有丰富的 B 族维生素，能够补益中气、温暖脾胃，对食欲不佳、脾胃虚寒、腹胀腹泻有着一定缓解作用。

马 拉 糕

❧

Name: Mara Cake

制作人：王志强　　中国烹饪大师

Made by: Zhiqiang Wang　　A Great Master of Chinese Cuisine

主　料　Main Ingredient
面粉：500g
flour　500g

配　料　Burdening
白糖：350g
suger　350g

鸡蛋：400g
egg　400g

调　料　Seasoning
泡打粉 Baking powder 25g
青红丝 Green red silk............20g
清油 oil 5ml

中国大锅菜

蒸烤箱卷（纪念版）

The Big-Wok-Made Cuisine of China, Food Volume of Steam Oven (Commemorative Edition)

制作方法

❶ 鸡蛋打入盆中打散，加入糖，继续搅拌，直到糖完全溶入蛋液中。

❷ 将面粉倒入蛋液中，搅拌均匀，加入泡打粉25g，边加水边搅拌，揉制成面团，面团较软，略稀。

❸ 将锡纸盒码入盘中，盒中刷少许底油，将面团用勺子舀入锡纸盒中，撒上青红丝，放入万能蒸烤箱蒸制成熟，选择「单点分层蒸煮」模式，时长20分钟，面点即成。

WORKING PROCESS

1. Put eggs into a mixing bowl, add sugar, stirring, until sugar is completely dissolved in the egg.

2. Add flour into the egg, stirring evenly, add baking powder, adding water while stirring, knead to make a dough. The dough is soft, slightly thin.

3. Will the tin box yards into the disk, brush a little oil on the bottom of the box, the dough with a spoon to scoop into the tin box, sprinkle with green and red silk, into the universal steam oven steaming mature, choose the mode of "single point layered cooking", 20 minutes, pastry is.

中国大锅菜

菜品名称·马拉糕
Name: Mara Cake

菜品特点

特色 马拉糕是传统的广式茶楼点心，质地松软，有海绵的柔软感，并带有鸡蛋的香味。传统做法中需要加入猪油，面粉发酵3日，如此则更为松软。本菜谱为适应团餐需要，采取简易做法，由面粉、鸡蛋、白糖制作而成，大大节省工序和制作时间。在制作过程中，泡打粉起到了关键关键作用，用量标准要严格遵守。

营养价值 马拉糕的原材料很简单，面粉、白糖、鸡蛋，它能有效补充人体所需能量，此外还有健脑、补心养肝和除热止渴的功效。

菜品名称

层酥烧饼

Name: Pastry Clay Oven Rolls

制作人：王志强　　中国烹饪大师

Made by: Zhiqiang Wang　　A Great Master of Chinese Cuisine

主 料 Main Ingredient	配 料 Burdening	调 料 Seasoning	
面 粉：500g	起酥油：250g	泡打粉 Baking Powder	5g
Flour　500g	Shortening　250g	酵 母 Yeast	5g
		芝 麻 Sesame	10g
		白 糖 Suger	40g

中国大锅菜

菜品名称·层酥烧饼
Name: Pastry Clay Oven Rolls

制作方法

① 将面粉倒在面板上，开成窝状，窝中倒入白糖40g、酵母5g、泡打粉5g，与面粉混合，加水和面。

② 将和好的面团用压面机压实，然后擀成大面片，将起酥油在面片上刷匀，然后撒上一层薄面。

③ 将面片卷起来成长条状，揪成剂子，团成椭圆的饼状，一面蘸满芝麻，另一面刷一层蛋液，放入万能蒸烤箱醒发，选择『发酵』模式，时长10分钟。

④ 将醒发好的烧饼放入万能蒸烤箱烤制成熟，选择『干烤曲奇—3号色』，时长12分钟，面点即成。

WORKING PROCESS

1. Put the flour on the panel, add sugar 40g, yeast 5g, baking powder 5g, and mixed with flour, add water to knead dough.

2. Press the dough with pressure surface compactor, and then roll into a large patches, Sorub the shortening on the surface evenly, then sprinkle with a layer of thin surface.

3. Roll The dough into strip, pulling it into one each small dough, mission into ellipse pie, dipsome sesame on one side, brush a layer of egg on the other side, put them into the Universal Steam Oven, choose the mode of "Fermentation" for 10 minutes.

4. Proofing the biscuits into the Universal Steam Oven, select "Dry Baking Cookies, color No.3" for 12 minutes.

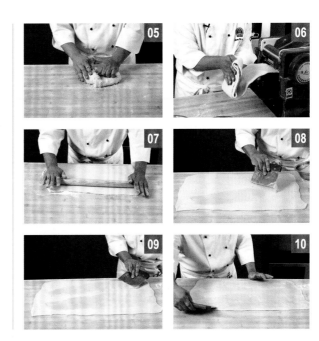

中国大锅菜

The Big-Wok-Made Cuisine of China, Food Volume of Steam Oven (Commemorative Edition)

蒸烤箱卷（纪念版）

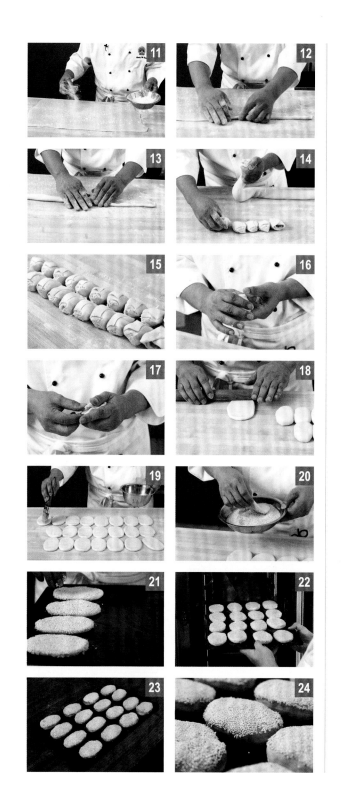

菜品特点

特色 层酥烧饼是北方地区最为常见的面点主食之一，以色泽金黄、芝麻香气浓郁、酥脆可口著称。制作过程中，成型后需要经过一定时间的醒发，然后方可进行烤制。

营养价值 面粉被人体吸收，转化为糖，提供日常活动所需能量。中医上讲，芝麻味甘、性平，能补肝肾、益精血、润肠燥，此外芝麻含大量脂肪油，又含有丰富的维生素B族，适合长期素食者。

菜品名称

蝴 蝶 酥

Name: Butterfly Cracker

制作人：王志强　　中国烹饪大师

Made by: Zhiqiang Wang　　A Great Master of Chinese Cuisine

主 料　Main Ingredient	猪　油：200g
面　粉：500g	pork oil　200g
Flour　500g	
	调 料　Seasoning
配 料　Burdening	白 糖　White Suger............200g
豆沙馅：200g	泡打粉　Baking Powder 5g
Red Bean Paste　200g	鸡 蛋　Egg................4 个 (pcs)

中国大锅菜

The Big-Wok-Made Cuisine of China, Food Volume of Steam Oven (Commemorative Edition)

蒸烤箱卷（纪念版）

制作方法

❶ 将面粉开成窝，中间倒入猪油 200g，白糖 200g，打入鸡蛋 4 个，揉制均匀，将白糖揉化。

❷ 面粉中加入泡打粉 5g，同面窝中间的配料一起揉成团，裹上保鲜膜醒发 15 分钟。

❸ 将醒发好的面团擀成大片，用钢圈抠出一个个圆面皮。

❹ 将豆沙馅擀成长条，切成一个个剂子，包入圆面皮中，用刀切开两个口，制成蝴蝶的形状，摆入烤盘，刷上鸡蛋液。然后放入万能蒸烤箱烤制成熟，选择『干烤曲奇—3 号色』模式，时长 6 分 30 秒，面点即成。

WORKING PROCESS

1. Put the flour into the nest, the add lard 200g, 200g sugar, 4 eggs in the middle of the flour.

2. Add 5g baking powder into wheat flour and knead it with the ingredients in dough, and then cover the dough with a preservative film to leaven ti for 15minutes.

3. Roll the leavened dough to the thin sheet, and the make round wrappers in it with steel rings.

4. Roll the bean-paste filling into a strip, then cut it into small pieces of filling and wrap the pieces in the round wrappers. Cut two openings in each dumpling and shape it as a butterfly, and then put them in a baking tray with egg wash brushed on them. Afterwards, bake them in the multi-function steaming oven in the "cookie toasting-color 3" mode for 6.5 minutes.

中国大锅菜

菜品名称·蝴蝶酥

Name: Butterfly Cracker

菜品特点

特色 蝴蝶酥以外形酷似蝴蝶而得名，外皮酥脆，馅料香甜，有点像夹心饼干，深受孩童喜爱。面点对于我们来说，既是一道主食，又是一件工艺品，大师只需简单的几下动作，就能勾勒出惟妙惟肖的外观。而美食制作背后，往往积淀着多年的功力，能控制好和面的手法、力度。此技艺则没有捷径，需要不断地苦练。

营养价值 蝴蝶酥味道香甜可口，可以想见其中少不了猪油、糖、鸡蛋等，这些食品均营养价值丰富，可以提供人体日常活动所需的蛋白、脂肪、糖分等；而红豆则具有清心养神、健脾益肾功效，此外红豆还具有帮助排水、消肿、轻身的功效。